UNDERGROUND

A

HUMAN HISTORY

OF THE

WORLDS BENEATH

OUR FEET

UNDERGROUND

Will Hunt

SPIEGEL & GRAU

NEW YORK

Published in the United States by Spiegel & Grau,
an imprint of Random House, a division of
Penguin Random House LLC, New York.

SPIEGEL & GRAU and colophon is a registered
trademark of Penguin Random House LLC.

LIBRARY OF CONGRESS CATALOGING-IN-PUBLICATION DATA
NAMES: Hunt, Will (Urban adventurer), author.
TITLE: Underground : a human history of the worlds
beneath our feet / by Will Hunt.
DESCRIPTION: First edition. | New York : Spiegel & Grau, [2018]
IDENTIFIERS: LCCN 2018005540 | ISBN 9780812996746
(hardback) | ISBN 9780812996753 (ebook)
SUBJECTS: LCSH: Underground construction—History. | Waste disposal
in the ground—History. | Mines and mineral resources—History. |
Burial—History.
CLASSIFICATION: LCC TA712 .H86 2018 | DDC 624.1/909—dc23 LC
record available at https://lccn.loc.gov/2018005540

Printed in the United States of America on acid-free paper

spiegelandgrau.com
randomhousebooks.com

246897531

First Edition

Book design by Barbara M. Bachman

To my parents

Nature loves to hide.

—HERACLITUS, *Fragments*

CONTENTS

═══

UNDERGROUND

DESCEND

There is another world, but it is in this one.

—PAUL ÉLUARD

Find signs of it everywhere you go. Step out your front door and feel beneath your feet the thrum of subway tunnels and electric cables, mossy aqueducts and pneumatic tubes, all interweaving and overlapping like threads in a great loom. At the end of a quiet street, find vapor streaming out of a ventilation grate, which may rise from a hidden tunnel where outcasts dwell in jerry-rigged shanties, or from a clandestine bunker with dense concrete walls, where the elite will flee to escape the end of days. On a long stroll through quiet pastureland, run your hands over a grassy mound that may conceal the tomb of an ancient tribal queen, or the buried fossil of a prehistoric beast with a long snaking spine. Hike down a shaded forest trail, where you cup your ear to the earth and hear the scuttle of ants excavating a buried metropolis, spoked with tiny whorled passageways. Trekking up in the foothills, you smell an earthy aroma emanating from a slender crack, the sign of a giant hidden cave, where the stony walls are graced with ancient charcoal paintings. And everywhere you go, beneath every step, you feel a quiver coming up from deep, deep below, where titanic bodies of stone shift and grind against one another, causing the planet to tremble and shudder.

If the surface of the earth were transparent, we'd spend days on our bellies, peering down into this marvelous layered terrain. But for us surface dwellers, going about our lives in the sunlit world, the underground has always been invisible. Our word for the underworld, *Hell,* is rooted in the Proto-Indo-European *kel-,* for "conceal"; in ancient Greek, *Hades* translates to "the unseen one." Today, we have newfangled devices—ground-penetrating radar and magnetometers—to help us visualize the underground, but even our sharpest images come out distant and foggy, leaving us like Dante, squinting into the depths: "so dark and deep and nebulous it was, / try as I might to force my sight below / I could not see the shape of anything." In its obscurity, the underground is our planet's most abstract landscape, always more metaphor than space. When we describe something as "underground"—an illicit economy, a secret rave, an undiscovered artist—we are typically describing not a place but a *feeling:* something forbidden, unspoken, or otherwise beyond the known and ordinary.

As visual creatures—our eyes, wrote Diane Ackerman, are the "great monopolists of our senses"—we forget about the underground. We are surface chauvinists. Our most celebrated explorers venture out and up: we have skipped across the moon, guided rovers into Martian volcanoes, and charted electromagnetic storms in distant outer space. Inner space has never been so accessible. Geologists believe that more than half of the world's caves are undiscovered, buried deep in impenetrable crust. The journey from where we now sit to the center of the earth is equal to a trip from New York to Paris, and yet the planet's core is a black box, a place whose existence we accept on faith. The deepest we've burrowed underground is the Kola borehole in the Russian Arctic, which reaches 7.6 miles deep—less than one half of one percent of the way to the center of the earth. The underground is our ghost landscape, unfolding everywhere beneath our feet, always out of view.

But as a boy, I knew that the underworld was not *always* invisible—to certain people, it could be revealed. In my parents' old edition of *D'Aulaires' Book of Greek Myths*, I read of Odysseus, Hercules, Orpheus, and other heroes who ventured down through craggy portals in the earth, crossed the river Styx on Charon's ferry, gave the slip to three-headed Cerberus, and entered Hades, the land of shades. The one who most captivated me was Hermes, the messenger god, he of the winged helmet and sandals. Hermes was the god of boundaries and thresholds, and the guide of the souls of the dead into Hades. (He bore the marvelous title of *psychopomp*, which means "soul conductor.") While other gods and mortals obeyed the cosmic boundaries, he swooped openly between light and dark, above and below. Hermes—who would become the patron saint of my own underground excursions—was the one true subterranean explorer, who cut through darkness with clarity and grace, who *saw* the underworld, and knew how to retrieve its buried wisdom.

The summer I turned sixteen, when my world felt as small and known as the tip of my finger, I discovered an abandoned train tunnel running beneath my neighborhood in Providence, Rhode Island. I'd heard about it first from a science teacher at school: a small, whiskery man named Otter, who knew every secret groove in every landscape in New England. The tunnel had once served a small cargo line, he'd told me, but that was years ago. Now it was a ruin: full of mud and garbage and stale air and who knew what else.

One afternoon, I found the entrance, which was concealed under a thicket of bushes behind a dentist's office. It was wreathed in vines and had the date of its construction—1908—engraved in the concrete above its mouth. The city had sealed the entrance with a metal gate, but someone had sliced open a small passageway: along with a few friends, I climbed underground, our flashlight beams crisscrossing in the dark. The mud sucked at our shoes and the air was boggy. On the ceiling were clusters of pearly, nipplelike stalactites that dripped water down on our heads. Halfway through, we dared one another to switch off our lights. As the tunnel fell to perfect darkness, my friends whooped, testing the echo, but I held my breath and stood dead still, as though, if I moved, I might float right off the ground. That night at home, I pulled up an old map of Providence. I started with my finger where we'd entered the tunnel and moved to where it opened at the other end. I blinked: the tunnel passed almost directly beneath my house.

That summer, on days when no one else was around, I'd put on boots and go walking in the tunnel. I couldn't have explained what drew me down, and I certainly never went with any particular mission. I'd look at the graffiti or kick around old bottles of malt liquor. Sometimes I'd turn off my light, just to see how long I could last in the dark before my nerves started to bristle. To the extent that I was aware of myself at all, as a sixteen-year-old I sensed that these walks were outside of my character: I was an uncertain teenager, scrawny, gap-toothed, with librarianish glasses. When my friends were starting to make out with girls, I still had a terrarium of pet tree frogs in my bedroom. I read about other people's adventures in books, without ever thinking to embark on my own. But something about the tunnel got under my skin: I'd lie in bed nights just imagining it running under the street.

At the end of that summer, following a heavy rainstorm,

I had just climbed beyond the threshold of the tunnel when I heard an unexpected rumble coming from the darkness ahead. Alarmed, I started to turn back, but decided to keep walking, even as the sound grew. Deep in the tunnel, I found the source: a crack in the ceiling—a burst pipe, maybe, or a leak—from which water was pouring down in cascades. Directly beneath the falling water, I saw an over-turned plastic beach pail. Then a paint bucket. Then, all at once, an enormous assembly of overturned containers—oil drums, beer cans, Tupperware, gas canisters, coffee tins—all in a giant cluster, arranged under mysterious circum-stances by a person I'd never meet. The water drummed down on the vessels, sending up an echoing song, as I stood in the dark, nailed to the floor.

Years passed, and I forgot about those underground walks. I left Providence, went to college, moved on. But my old connection to the tunnel never quite disappeared. Just as a seed silently takes root and ripens and matures down in the hidden earth before sprouting on the surface, my mem-ories of the tunnel germinated for years at the bottom of my mind. It was not until much later, following a series of unexpected encounters with the underground of New York City, that my old memories of the tunnel, and of the mys-terious altar of buckets, reemerged. When they did, they seized me with a ferocity that turned my entire imagina-tion inside out, fundamentally altering the way I thought about myself, and my place in the architecture of the world.

I came to love the underground for its silence and for its echoes. I loved that even the briefest trip into a tunnel or a cave felt like an escape into a parallel reality, the way characters in children's books vanish through portals into secret worlds. I loved that the underground offered romping Tom Sawyer–ish adventure, just as it opened confrontation with humankind's most eternal and elemental fears. I loved telling stories about the underground—about relics discovered beneath city streets, or rituals conducted in the depths of caves—and the wonder they engendered in the eyes of my friends. Most of all, I was captivated by the dreamers, visionaries, and eccentrics that the underground attracted: people who'd heard a kind of siren song and given themselves to exploring, making art, or praying in the underground world. People who had surrendered to obsession in a way I thought I understood, or at least wanted to understand. Down in the dark, I thought I might find a perverse kind of enlightenment.

Over the years, I convinced a research foundation, then various magazines, then a book publisher to give me funding to investigate these things, and I spent that money, and then some, exploring subterranean spaces in different parts of the world. For more than a decade, I climbed down into stony catacombs and derelict subway stations, sacred caves, and nuclear bunkers. It began as a quest to understand my own preoccupation; but with each descent, as I became attuned to the resonances of the subterranean landscape, a more universal story emerged. I saw that we—all of us, the human species—have always felt a quiet pull from the underground, that we are as connected to this realm as we are to our own shadows. From when our ancestors first told stories about the landscapes they inhabited, caves and other spaces beneath our feet have frightened and enchanted us, forged our nightmares and fantasies. Underground worlds, I discovered, run through our history like a secret thread: in ways subtle and profound, they have guided how we think about ourselves and given shape to our humanity.

THE UNDERGROUND OPENED SLOWLY, in small, splintering cracks—then all at once, like a trapdoor beneath my feet. It began the first summer I lived in New York, when I was working at a magazine in Manhattan and living in Brooklyn with my aunt and uncle and my two cousins, Russell and Gus. After years spent as a teenager, envisioning myself as a future New Yorker who would take long, ecstatic night walks through Manhattan, absorbing every dot of light emanating from every apartment window, I had arrived to find the city impenetrable. I shrank in crowds, stammered to bodega keepers, and got off at the wrong subway stop, only to wander through Brooklyn, feeling like a hayseed, too embarrassed to ask for directions.

Late one night, when I was feeling particularly cowed by the city, I was waiting for the subway in lower Manhattan, on one of the deep-set platforms where on summer nights you can almost smell the city's granite bedrock, when I saw something that baffled me. From the darkness of the tunnel materialized two young men: they wore headlamps and their faces and hands were black with soot, as though they'd been climbing for days through a deep cave. They fast-walked up the tracks, clambered onto the platform right in front of me, and vanished up the stairs. I rode the train home that night with my forehead pressed against the window, fogging the glass, imagining a whole secret honeycomb of spaces hidden beneath the streets.

The young men in headlamps were urban explorers, part of a loose confederacy of New Yorkers who'd made a pastime of infiltrating the off-limits and secreted-away spaces under the surface of the city. They were a kingdom of many tribes: some were historians who documented the grandeur of the city's forgotten places; others, activists who trespassed to symbolically reclaim New York's corporatized spaces; still others were artists, who assembled secret installations and staged performances in the city's obscure layers. In those first weeks, as I puzzled my way through New York, I found myself staying up late at night, studying explorers' photographs of hidden places—subway stations abandoned for decades; deep valve chambers in the water system; derelict bomb shelters coated in dust—all of which felt as exotic and mysterious as lost sea monsters gliding through the deep ocean.

One night, while sifting through an explorer's archives, I was startled to find myself looking at a photograph of the tunnel I'd explored as a boy in Providence. I hadn't thought about it in years: the single rail receding into the dark, the "1908" carved above the entrance. The intimacy of this chance encounter was almost disquieting, as though someone had reached down into my mind and opened a hatch, allowing a whole raft of buried memories to float to the surface. The photographer, I learned, was a man named Steve Duncan: a dashing and brilliant and possibly deranged individual who would become my first guide into the underground.

We met one afternoon on a scouting trip to the Bronx, where Steve was plotting an excursion through an old sewer pipe. He was six or seven years older than me, sandy-haired and blue-eyed, with the rangy build of a rock climber. He'd begun exploring during his freshman year at Columbia University, where he would go slinking through a network of steam tunnels under the campus. One night, he wriggled through a vent in a wall and emerged into a chamber cluttered with moldering scientific equipment. It was the storeroom for the earliest incarnation of what would eventually become the Manhattan Project. The bulbous green machine at the center of the room was the original particle accelerator: a strange jewel of history, hidden just out of view.

A fascination took hold, and Steve promptly switched his major from engineering to urban history. When he wasn't studying, he began exploring train tunnels, then strapped on waders and sloshed through sewers, and before long began climbing to the tops of suspension bridges, where he took soaring, omniscient photographs of the city. Over the years, he styled himself as a guerrilla historian and photographer, with an alarmingly granular knowledge of the city's infrastructure. (The Department of Environmental Protection, which monitored the city's sewers, periodically tried to hire Steve, despite his illegal research methods.) Steve was somewhere between nerd and outlaw: he was skinny, had traces of a boyhood speech impediment, but drank like a longshoreman, had a raffish smile that women loved, and carried himself with a heroic strut. As a young man, he'd contracted a rare form of bone cancer in his leg and had nearly lost his life, an experience that seemed to infuse everything he did with urgency and vitality. He could spend a night enumerating the significances of the various acronyms embossed on the city's manhole lids, or expounding upon changes in flow rates of nineteenth-century European wastewater systems, then go out and get into a bar fight.

That afternoon, we zigzagged between sewer grates, shining flashlights underground, tracing the route of the pipe. As we went, Steve talked about the city's jigsaw of hidden systems with a kind of evangelical love. He envisioned New York as a gigantic, shifting, many-tentacled organism, of which surface dwellers saw only a sliver. It was his mission to reconnect people with the hidden aspect of the world: he wished every manhole cover in every city were made of glass, allowing people to peer underground at any moment.

"Most people move through the world in two dimensions," he said. "They have no idea what's beneath them. When you see what's underground, you understand how the city works. But it's more than that. You see your place within history, how you fit in the world."

In Steve, I found an incarnation of Hermes, who could see a parallel topography. "I believe that much unseen is also here," wrote Walt Whitman in *Leaves of Grass*. Steve saw what was unseen—I wanted to see it, too.

MY FIRST DESCENT WAS a gentle one: a walk through the West Side Tunnel—known to explorers and graffiti writers as the Freedom Tunnel—which runs for some two and a half miles beneath Riverside Park on the Upper West Side of Manhattan.

On a summer morning, I slipped through a slice in a chain-link fence near 125th Street and headed through the tunnel's entrance, which was voluminous, maybe twenty feet high and twice that across. Rather than pitch dark, the tunnel was twilit: every few hundred feet there was a rectangular ventilation grate in the ceiling, which, like a cathedral window, let in soft columns of light. I set off on a quiet walk through the middle of Manhattan without seeing another soul, like something from a dream.

Around the halfway point, I came upon an enormous mural, more than one hundred feet long, painted by an art-

ist named Freedom, who was the tunnel's namesake. I stood by the opposite wall, admiring the painting, which seemed to tremble in the light. A soft breeze blew through, and I could hear the faraway clamor of cars on the West Side Highway mingling with birdcalls in the park.

And then, from down the tunnel, I saw the giant headlight of a train moving toward me. Crouching with my back against the wall, I felt a deep bass tremor under my feet, and then, all at once, a blast of light, a gale-force wind, and a roar that sent a vibration down through my ribs. I had not been in any real danger—there was a good fifteen feet of clearance between me and the tracks—but as I crouched there my whole body quaked, my mind on fire.

Even as I emerged from the tunnel that first afternoon, climbing over a fence near the Hudson River, my relationship with the city had begun to shift. Aboveground, I traced and retraced a single path between work and home, following a narrow track of sensory experience; down in the tunnel, I stepped outside those bounds, and connected to the city in a new and visceral way. I felt shaken awake, like I was making eye contact with New York for the first time.

Going underground, crawling down into the body of the city, became my way of proving to myself that I belonged in New York, that I *knew* the city. I liked being able to tell my born-and-raised-Manhattanite friends about old vaults beneath their neighborhood that they knew nothing about. Down in sunken alleys, I enjoyed seeing textures of the city that were invisible to people on the surface— ancient graffiti tags, cracks in the foundations of skyscrapers, exotic molds creeping over walls, decades-old newspapers crumpled in hidden crevices. New York and I shared secrets: I was sifting through hidden drawers, reading private letters.

Near the Brooklyn Navy Yard one night, Steve laid orange construction cones around a manhole and popped open the lid with an iron hook, releasing a twist of vapor from below. We climbed down the ladder, hand over hand on slimy rungs, and splashed down in a sewer collector: it was maybe twelve feet high, with greenish water burbling down the center. The air was warm; my glasses immediately fogged. I hesitated before long, gooey strings of bacteria dangling from the ceiling—affectionately known as "snotsicles"—but the sewer was less repellent than I'd expected. The smell was not so much fecal as it was earthy, like an old farm shed full of fertilizer. We passed our lights over banks of loamy muck, like sandbars in a river, where tiny crops of albino mushrooms grew. During migration season, eels swam through this water.

Mixed in with the green flow, Steve told me, was Wallabout Creek, an old waterway that filtered into Wallabout Bay, where the Navy Yard was now located. You could see the creek on a 1766 map, but as the city had grown and expanded, it had been forced underground and out of sight.

Up on the surface, I knew New York as a crude, if exuberant, animal: rumbling and growling, belching steam and disgorging crowds from its various orifices. But down here, with one of its ancient streams flowing quietly around my feet, the city felt serene, even vulnerable. It was intimate almost to the point of embarrassment, the feeling of watching someone as they slept.

It was past three in the morning when we climbed back up the ladder and emerged through the open manhole into bracing cold air. Just as we came up, a young man on a bike swerved to avoid us. He skidded and spun back around. Breathlessly, he asked, "Who are you guys?"

Steve pulled himself straight and puffed out his chest, as though he were standing onstage, then tilted back his head and delivered lines from the Robert Frost poem "A Brook in the City":

> *"The brook was thrown*
> *Deep in a sewer dungeon under stone*
> *In fetid darkness still to live and run—*
> *And all for nothing it had ever done,*
> *Except forget to go in fear perhaps."*

With each trip underground, the city cracked open a little bit more, disclosing another secret, just enough to draw me deeper. I rode the subway with a notebook, watching out the window and recording the locations of gaps in the walls that would potentially lead to abandoned platforms, or "ghost stations," as graffiti writers call them. I tracked routes of underground streams and searched for places where you could hold your ear to a grate and hear water babbling beneath the surface. My closet hung with waders and mud-soaked clothing, and I carried a headlamp in my backpack at all times. I began moving through the city more and more slowly, as I paused to peer into subway vents, sewer manholes, and construction pits, trying to puzzle together the city's innards. My mental map of New York came to resemble a coral reef, riddled with hidden creases, secret passages, and unseen pockets.

For a while, I moved through the city in a kind of delirium, imagining that every manhole, every dark stairwell and hatch in the street was a portal to another layer. I discovered a brownstone in Brooklyn Heights that looked like any other house on the block, except its door was made of industrial steel and its windows were blacked out: it was a camouflaged ventilation shaft leading down into the subway. On Jersey Street in SoHo, I found an old manhole that opened down into an antiquated water tunnel called the Croton Aqueduct, where, in 1842, four men piloted a small wooden raft called the *Croton Maid* on a forty-mile subterranean odyssey in pitch darkness from the Catskills to Manhattan. Beneath Atlantic Avenue in Brooklyn, I visited a train tunnel that had been abandoned in 1862, and forgotten by the city until 1980, when a nineteen-year-old local man named Bob Diamond climbed down through a manhole and discovered the gigantic, echoing hollow. (The discovery cast the city into a brief, seething fascination, as photographers rushed to capture images of young Bob crawling through the lost tunnel.) On an island in the Bronx, I joined a team of treasure-hunters in the search for a bundle of lost ransom money allegedly buried there by the man who kidnapped Charles Lindbergh's baby. I followed rumors of certain sepulchral chambers hidden in the subway system where the walls were inscribed with century-old graffiti: spaces so untouched and forgotten that if you raised your voice, sand would fall in cascades from the ceiling. I searched for an old building in Midtown where the sub-basement reportedly had a hole in the floor that opened down onto a flowing river, around which old men would sit during the day, fishing for trout. I told these stories so often that my friends tired of them, just as they groaned when, on walks through the city, I tried to narrate what was beneath our feet at any given moment, but by this point, I couldn't help myself.

Down I climbed, taking risks that surprised even myself. Late at night, I'd step over the DO NOT ENTER OR CROSS TRACKS sign at the end of a subway platform, slink along the catwalk, then hop down onto the tracks, where it was chimney-black and, on summer nights, as hot as a furnace. At first, I went down with my cousin Russell: half-running in the dark, we'd feel a faint stirring of air, then a subsonic tremble underfoot. "Train coming," we'd whisper. We'd hear the tracks lock into place, then, from around the bend, we'd see a giant headlight illuminating the tunnel wall: we'd scramble up onto the catwalk, then fold ourselves into an emergency-exit alcove as the train roared past, blowing a gust of wind so strong it could knock you over. Soon I started going down alone, on impulsive, spur-of-the-moment excursions. Waiting on a platform late at night after a party, or a long night in the library, I'd see the train coming, but at the last second, I'd decide to let it pass,

only to follow it down the tunnel into the dark. I had close calls and brushes, where I saw blue sparks flying from passing wheels, where the roar of a train would leave me temporarily deaf. Late at night, I'd walk home in a daze, my cheeks smudged with steel dust, feeling both inside and outside of myself, as though having just awoken from a dream.

One night a train conductor spotted me in the tunnels, and I emerged onto a platform to find two NYPD officers waiting for me: two young Dominican guys, one short and fat, one tall and skinny. They pinned my arms, pushed me against a wall, and emptied my backpack onto the floor— despite having every reason to arrest me, they ultimately let me go. I sat on a bench up on the street, feeling rattled and foolish, and fully aware that if I weren't a white man, I'd be in cuffs. Even that night, as I walked home, I found myself pausing in the street, peering into grates and manholes.

In the darkest strata I encountered the Mole People, the homeless men and women who lived in the city's hidden nooks and vaults. One night, along with Steve, Russell, and a few other urban explorers, I met a woman named Brooklyn, who'd been living underground for thirty years. She had a pocked face, with dreadlocks piled high on her head. Her home, which she called her "igloo," was an alcove hidden in the eaves of the tunnel, furnished with a mattress and a few crooked pieces of furniture. It was Brooklyn's birthday: we passed around a bottle of whiskey, as she sang a medley of Tina Turner and Michael Jackson songs, and for a while everyone laughed. But then something came undone, Brooklyn's singing gave way to glossolalia, and she started seeing things that weren't there. Her partner returned—he was also named Brooklyn, somehow—and the two of them fought, shrieking in the dark.

Over time, I stopped talking about my underground trips with friends and family, finding it more difficult to answer their questions: what was I searching for down there?

"I'M GOING TO SHOW you something," said Steve one night, "but you have to promise not to say a word about it to anyone."

It was around two, we were leaving a party somewhere in Brooklyn. Steve led me into a subway station, then down the catwalk, where I followed close behind, until he suddenly vanished in the dark. Only when I heard his voice did I realize that he had slipped through a hidden portal in the wall. I stepped through and emerged into a dark void. It was one of the hallowed spaces of New York's underground, a giant, echoing cavity, separated from ordinary life by the thinnest of membranes, and yet completely unseen.

Steve led me to the middle of the room and shined his light down at the floor. It was a rectangular grid of ceramic tiles, maybe six by four feet, blanketed in dust. We blew on the tiles, sending up a cloud. I found myself looking at a map. It was a rendering of the New York subway map, the one you see on the walls of every station in the city, with the lumpy, beige silhouettes of Brooklyn and Manhattan, and the train lines snaking across the pale blue East River. But rather than the familiar landmarks, the map showed only the invisible places. New York's veteran urban explorers, who'd spent years infiltrating the city, had affixed photographs to the map, each showing the location of a sewer, an aqueduct, a ghost station, or another place excluded from public view. As I crouched in the dark, studying this atlas of the invisible city, itself hidden in an invisible place, I was exhilarated: here was a kind of shrine to everything I'd sought in years of exploring under New York. And yet, at the same time, I felt oddly estranged from that exhilaration, as though it were buzzing up from a part of my mind not entirely familiar to me.

In that moment, as I stood in the dust deep under the city, I began to see just how little I understood about my connection to the underground. And for that matter, how little I understood about humankind's larger relationship to this landscape, which reaches back into the faintest glimmerings of our history.

Once, on a hike in Tuscany, Leonardo da Vinci ambled over a span of boulders and came upon the mouth of a cave. As he stood in the shadow of the threshold, feeling a cool breeze on his face, he stared into the dark, and found himself at an impasse. "Two contrary emotions arose in me," he later wrote, "fear and desire—fear of the threatening, dark cavern, desire to see whether there were any marvelous things in it."

We have lived among caves and underground hollows for as long as humans have existed, and for just as long, these spaces have evoked in us visceral and perplexing emotions. Evolutionary psychologists have suggested that even our most archaic ancestral relationships to landscapes never quite fade, that they become wired in our nervous system, manifest in unconscious instincts that continue to govern our behavior. The ecologist Gordon Orians calls these lingering vestigial impulses "evolutionary ghosts of environments past." On my trips under New York, each time I peered into the mouth of a dark tunnel or down into a sewer manhole, I was unconsciously sifting through ghost impulses inherited from ancestors, who long ago crouched at the mouths of dark caves, deciding whether or not to descend.

We are aliens underground. Natural selection has designed us—in every way imaginable, from our metabolic needs, to the lattice-like anatomy of our eyes, to the deep, jellied structures of our brain—to stay on the surface, to *not* go underground. The "dark zone" of a cave—scientists' name for the parts of a cave beyond the "twilight zone," which is within the reach of diffuse light—is nature's haunted house, a repository of our deepest-rooted fears. It is home to snakes that twitch down from cave ceilings, spiders the size of Chihuahuas, scorpions with barbed tails—creatures we are evolutionarily hardwired to fear, because they so often killed our ancestors. Up until about fifteen thousand years ago, caves all over the world were the dwelling place of cave bears and cave lions and saber-toothed tigers. Which is to say, for all but the last eyeblink of our species' existence, every time we came upon a cave mouth, we braced ourselves for a man-eating monster to lunge out of the darkness. Even today, when we peer underground, we feel the flickering dread of predators in the dark.

As we evolved for life on the African savannah, where we hunted and foraged in daylight and where nocturnal predators stalked us in the night, darkness has always unnerved us. But subterranean dark—in "the sightless world," as Dante called it—is enough to cause our entire nervous system to splinter. The pioneering cave explorers of modern Europe imagined that a prolonged stay in subterranean darkness could permanently dismantle their psyches, as one seventeenth-century writer described upon exploring a cave in Somerset, England. "We began to be afraid to visit it," he wrote, "for although we entered in frolicksome and merry, yet we might return out of it sad and pensive and never more to be seen to laugh whilst we lived in the world." Which, in some ways, has turned out to be true, as neuroscientists have demonstrated manifold ways in which prolonged immersion in absolute darkness can trigger psychological aberrations. In the 1980s, on an expedition into a cave called Sarawak Chamber in the Mulu National Park in Borneo, a caver entered a gigantic cavern—big enough to fit seventeen football fields, one of the largest cave rooms in the world—and lost track of the rock wall: as he drifted through interminable darkness, the caver fell into a kind of paralytic shock, and had to be guided out by his partners. Cavers have referred to such darkness-fueled panic attacks as "the Rapture."

The feeling of enclosure, too, leaves us unhinged. To be trapped in an underground chamber, our limbs restricted, cut off from light, with oxygen dwindling, may be the emperor of all nightmares. The ancient Roman philosopher Seneca once described a crew of silver prospectors climbing deep into the earth, where they encountered phenomena "fit to make them shake with horror," among which was the psychic pressure of a mass of "land hanging above their heads." It was a sentiment echoed by Edgar Allan Poe, the poet laureate of claustrophobia, who wrote of subterranean enclosure: "No event is so terribly well adapted to inspire the supremeness of bodily and of mental distress. . . . The unendurable oppression of the lungs—the stifling fumes from the damp earth—the clinging to the death garments— the rigid embrace of the narrow house—the blackness of the absolute Night—the silence like a sea that over- whelms . . ." Down in any underground hollow, we feel, if not a full tempest of panic, a reflexive tingle of not-quite- rightness, as we imagine ceilings and walls closing in on us.

Ultimately, it's death we fear most: all of our aversions to the dark zone come together in the dread of our own mortality. Our species has been burying our dead in the dark zones of caves since at least one hundred thousand years ago, according to discoveries in Qafzeh Cave in Israel, and our Neanderthal ancestors since long before that. In religious traditions all over the world, descriptions of the realm of the dead mirror the dark zones of caves, where disembodied shades drift through edgeless darkness. Even cultures who occupy landscapes without caves, who don't come in contact with physical underground space—peoples of the Kalahari Desert or the Siberian flatlands—tell myths of a vertical cosmos, where a subterranean realm swarms with spirits. Each time we cross a cave threshold, we feel a reflexive premonition of our eventual death—which is to say, we brush up against the one thing natural selection has designed us to avoid.

And yet, when we crouch at the edge of the under- ground, we *do* descend. On that day in Tuscany, Leonardo da Vinci *did* climb down into the dark. (Embedded in the cave wall, deep in the dark zone, he discovered the fossil of a whale, which would haunt and inspire him for the rest of his life.) Virtually every accessible cave on the planet contains the footprints of our ancestors. Archaeologists have belly-crawled through muddy passageways in the caves of France, swum down long subterranean rivers in Belize, and trekked for miles inside the limestone caves of Kentucky: everywhere, they have found fossilized traces of ancient people, who clambered down through rocky apertures in the earth, lighting their way through the gloom with pine torches or fat-oil lamps. There our ancestors encountered an alien realm, completely severed from what they knew on the surface: a world darker than any night, where echoes boomed, where stalagmites, like monsters' teeth, spiked up from the floor. The dark zone journey may well be human-kind's oldest continuous cultural practice, with archaeo-logical evidence going back hundreds of thousands of years, before our species even existed. No single tradition, writes the mythologist Evans Lansing Smith, "brings us all to-gether as human beings more than the descent to the un-derworld."

And so as I began to dissect my own preoccupation with the underground of New York, I found myself wrapped up in a much larger and older and more universal mystery. Despite the most basic evolutionary logic, despite every immediate subterranean peril, despite a chorus of hard-wired fears urging us to stay in the light, despite even the visceral evocation of our own death, we possess an impulse, buried in the core of our psyches, that draws us into the dark.

OVER YEARS, I TRAVELED widely, yo-yoing between New York and remote corners of the world, following every thread in our tangled relationship to the subterranean landscape. From dank corridors beneath modern cities, I went into older and wilder hollows, and finally into the ancient darkness of natural caves. In each place, I was guided by another underground devotee, an incarnation of Hermes, who knew the underworld intimately, and traveled openly between above and below.

"To go down to the cellar is to dream," wrote the philosopher Gaston Bachelard in *The Poetics of Space*. "It is losing oneself in the distant corridors of an obscure etymology, looking for treasures that cannot be found in words." As I traced our relationship to the underground through mythology and history, art and anthropology, biology and neuroscience, I found a symbol bewildering in its expansiveness, a landscape as elementary to the human experience as water, air, or fire. We go underground to die, but also to be reborn, to emerge from the womb of the earth; we dread the underground, and yet it is our first refuge in times of danger; it conceals priceless treasures, alongside toxic waste; the underground is the realm of repressed memory and of luminous revelation. "The underground metaphor," wrote scholar David L. Pike in his book *Metropolis on the Styx*, "can be expanded to encompass all of life on earth."

To become conscious of the spaces beneath our feet is to feel the world unfold. As we turn our thoughts to the tunnels and caves in the physical underground, we become attuned to all of the invisible forces that shape our reality. Our connection to the underground cracks a door into the inscrutable chambers of the human imagination. We go down to see what is unseen, unseeable—we go in search of illumination that can be found only in the dark.

THE CROSSING

And I stared through that obscurity,
I saw what seemed a cluster of great towers,
Whereat I cried, "Master, what is this city?"

—DANTE, CANTO XXXI, *The Inferno*

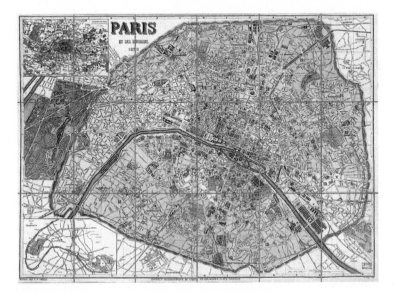

The first person to photograph the underground of Paris was a gallant and theatrical man with a blaze of red hair, known as Nadar. Once described by Charles Baudelaire as "the most amazing example of vitality," Nadar was among the most visible and electric personalities in mid-nineteenth-century Paris. He was a showman, a dandy, a ringleader of the bohemian art world, but he was known especially as the city's preeminent photographer. Working out of a palatial studio in the center of the city, Nadar was a pioneer of the medium, as well as a great innovator. In 1861, Nadar invented a battery-operated light, one of the first artificial lights in the history of photography. To show off the power of his "magic lantern," as he called it, he set out to take photographs in the darkest and most obscure spaces he could find: the sewers and catacombs beneath the city. Over the course of several months, he took hundreds of photographs in subterranean darkness, each requiring an exposure of eighteen minutes. The images were a revelation. Parisians had long known about the cat's cradle of tunnels, crypts, and aqueducts beneath their streets, but they had always been abstract spaces, whispered about, but seldom seen. For the first time, Nadar brought the underworld into full view, opening Paris's relationship to its subterranean landscape: a connection that, over time, grew stranger, more obsessive, and more intimate than that of perhaps any city in the world.

A century and a half after Nadar, I arrived in Paris, along

with Steve Duncan and a small crew of urban explorers, with an aim to investigate the city's relationship to its underground in a way no one had before. We planned a traverse—a walk from one edge of the city to the other, traveling exclusively by subterranean infrastructure. It was a trip Steve had dreamed up back in New York: we'd spent months planning, studying old maps of the city, consulting Parisian explorers, and tracing potential routes. The expedition, in theory, was tidy. We would descend into the catacombs just outside the southern frontier of the city, near Porte d'Orléans; if all went according to plan, we'd emerge from the sewers near Place de Clichy, beyond the northern border. As the crow flies, the route was about six miles, a stroll you could make between breakfast and lunch. But the subterranean route—as the worm inches, let's say—would be winding and messy and roundabout, with lots of zigzagging and backtracking. We had prepared for a two- or three-day trek, with nights camping underground.

On a mild June evening, six of us sat on the southern boundary of the city, in a derelict train tunnel that was part of the *petite ceinture,* or the "little belt," a long-abandoned train track that encircles Paris. We'd spent the day collecting last-minute supplies: now it was past nine, and the dots of light at either end of the tunnel were darkening. Everyone was quiet, our headlamp beams dancing anxiously over the floor. We took turns peering down into a dark, graffiti-ringed hole jack-hammered out of the concrete wall, which would be our entrance into the catacombs.

"Best to keep passports in a zipper pocket," said Steve, thumbing the braces on his waders. "Just in case." Every step of the trip, of course, would be illegal: if we got caught, having our IDs at the ready might be just enough to keep us out of Paris's central lockup.

Moe Gates crouched over a map that would help us navigate the sprawling, mazelike tunnels of the catacombs. Short, bearded, and clad in a red Hawaiian shirt, Moe was Steve's longtime exploring partner. He had run the sewers in Moscow, crouched on the gargoyles at the top of the Chrysler Building in Manhattan, and once had sex on the top of the Williamsburg Bridge in Brooklyn. He wanted to retire from exploring tunnels, to settle down and "have babies with a nice Jewish girl," but he hadn't been able to kick the habit.

Liz Rush—Steve's girlfriend, a sharp-eyed woman with chestnut hair cropped above her shoulders—was checking batteries on a confined-space gas detector, which would alert us to any poisonous air that we might encounter in the unventilated tunnels. Liz had explored under New York with Steve, but this was her first trip beneath Paris. Sorting through gear next to Liz were two other first-timers: Jazz Meyer, a young Australian woman with red dreadlocks, who had explored storm drains under Melbourne and Brisbane; and Chris Moffett, a philosophy graduate student in New York, who would be making his first foray underground.

"Fifty percent chance of precipitation," said Steve, checking his phone one last time before turning it off. The greatest threat to our trip was rain: once we reached the sewer collectors, even a small cloudburst on the surface could create a flood underground. It had been a wet June in

Paris and, since our arrival in the city, we'd been obsessively monitoring the weather. Steve had enlisted a fellow explorer in the city, Ian, to text us weather updates. As a group, we'd made a vow: at the first sign of raindrops, we'd bail, expedition over.

As we huddled around the entrance, Moe, who would play the role of recordkeeper, checked his watch and made a note on a pad: "Nine forty-six P.M., underground." Steve went first, snaking his hips through the entrance, legs scissor-kicking behind him; the rest of us followed, one after the next. I was last: I looked up and down the empty rail tunnel, took a deep breath, then squeezed down into the dark.

The tunnel we dropped into was narrow and low, with walls of raw, clammy stone. I slung my pack around to my chest and crawled on all fours, my back scraping against the rocky ceiling, while cold water sloshed around my hands and knees, soaking me to the skin. The stone gave off an earthy, almost pastoral aroma, like rain-soaked chalk. Our headlamp beams flitted in an arrhythmic strobe. So abrupt was the feeling of departure from the surface, we might as well have been at the bottom of the ocean. The honking of cars on the street, the rattle of the tram on avenue du Général Leclerc, the murmur of Parisians smoking under the awnings of brasseries—all were stamped out.

We headed north, with Steve in the lead. Down a wider gallery, we rose into a squelching duck-walk, then down an arched passage, with earthen ground underfoot, until all of us were up and marching, the first leg of our traverse under way.

Parisians say their city, with its galaxy of perforations, is like a great hunk of Swiss cheese, and nowhere is so holey as the catacombs. They are a vast, stony labyrinth, two hundred miles of tunnels, mainly on the Left Bank of the Seine. Some of the tunnels are flooded, half-collapsed, riddled with sinkholes; others are adorned with neatly mortared brick, elegant archways, and ornate spiral staircases. The "catas," as they are known to the familiar, are technically not catacombs, a word usually traced back to an amalgam of the Greek *katá-* (down) and Latin *tumbae* (tombs); they are quarries. All of the stately buildings along the Seine—Notre-Dame, the Louvre, the Palais Royal—were erected of limestone blocks chopped from beneath the city. The oldest tunnels had been carved to construct the Roman city of Lutetia, traces of which could still be found in the city's Latin Quarter. Over the centuries, as the city grew, stonecutters brought more limestone to the surface, and the underground warren expanded, fanning out beneath the city like the roots of a great tree.

In the years before Nadar first brought his camera beneath Paris, the quarries were silent. The only regular visitors were a handful of city laborers—the workers in the ossuary, who raked bones back and forth over the catacomb floors; the employees of the Inspection générale des carrières, who walked the stone passages by lantern light, bracing the tunnels to prevent them from collapsing under the city's weight—and the occasional mushroom farmer, who took advantage of the dry, dark environment to grow his crop. For the rest of the city, the quarries were a blind spot: a distant place, a landscape more imaginary than real.

From the moment we went underground, so many years after Nadar, we could feel the quarries teeming. The walls tumbled with bright graffiti, and the mud floors were tracked up and down with footprints. When we came to shallow pools, the water swirled with mud, a sign of recent passers-through. These were traces of the *cataphiles,* a loose affiliation of Parisians who spent days and nights roaming the catacombs. A subtribe in the urban explorer kingdom, cataphiles were mostly college kids in their teens and twenties; some, however, were in their fifties and sixties, had been exploring the network for decades, had even raised cataphile children and grandchildren. The city employed a squadron of catacomb policemen—known as *cataflics*, literally "catacops"—who patrolled the tunnels and doled out sixty-five-euro tickets to trespassers. But they offered little deterrence to the cataphiles, who treated the tunnels like a giant secret clubhouse.

We'd been underground for about two hours when Steve led us through a tunnel so tight and low we dropped to our bellies and squirmed on our elbows through the dirt. As we popped out on the other side, we saw three headlamp beams bobbing in the dark. It was three young Parisians—cataphiles—led by a tall, rangy, dark-haired man in his mid-twenties named Benoit.

"Welcome," he said, with a flourish, "to La Plage."

We'd emerged in one of the main cataphile haunts, a cavernous chamber with sand-packed floors and high ceilings supported by thick limestone columns. Every surface—every inch of the wall, of the pillars, and much of the rocky ceiling—was covered in paintings. In the darkness, the paintings were subdued and shadowy, but under the beam of a flashlight, they blazed. The centerpiece was a replica of Hokusai's *Great Wave off Kanagawa*, with the curling wave of frothy blues and whites. Spread throughout the room were stone-cut tables, rough-hewn benches and chairs. At the center of the chamber was a giant sculpture of a man with arms raised to the ceiling, like a subterranean Atlas, holding up the city.

"This is like—" Benoit paused, apparently searching for a recognizable analogy "—the Times Square of the catacombs."

On weekend nights, he explained, La Plage and certain other voluminous chambers in the catacombs filled with revelers. Sometimes they'd siphon electricity from a lamppost on the surface and set up a band or a DJ. Or a cataphile would strap a boom box to their chest and go weaving through the tunnels, roaming from chamber to chamber, as the party followed, dancing in the dark, passing bottles of whiskey up and down, like a snaking subterranean conga line. Other gatherings were more urbane: you might turn in to a dark chamber to find a candlelit holiday party, with cataphiles drinking champagne and eating *galette des rois*.

Cataphiles had long flocked underground to make art, to paint and sculpt and build installations in hidden caverns. Not far from La Plage was the Salon du Chateau, where a cataphile had carved from stone a beautiful replica of a Norman castle and installed gargoyle sculptures in the wall. And the Salon des Miroirs, where the walls of a chamber were covered in a disco-ball mosaic of reflective shards.

And La Librairie, a small nook with hand-carved shelves, where people could leave books for others to borrow. (The books, unfortunately, often grew moldy in the dank air.)

To wander through the catacombs is to feel yourself inside of a mystery novel, full of false walls and trapdoors and secret chutes, each leading to another hidden chamber, containing another surprise. Down one passageway, you might find a chamber containing a sprawling Boschian mural that cataphiles had been gradually embellishing for decades; down another, you might see a life-size sculpture of a man half inside a stone wall, as though stepping in from the beyond; down yet another, you might encounter a place that upends your very sense of reality. In 2004, a squadron of cataflics on patrol in the quarries broke through a false wall, entered a large, cavernous space, and blinked in disbelief. It was a movie theater. A group of cataphiles had installed stone-carved seating for twenty people, a large screen, and a projector, along with at least three phone lines. Adjacent to the screening room were a bar, lounge, workshop, and small dining room. Three days later, when the police returned to investigate, they found the equipment dismantled, the space bare, except for a note: "Do not try to find us."

Whether or not they knew it, the cataphiles were essential to our traverse. Our map, which had been designed by the tribe's elders, was a product of generations of cataphile knowledge: it marked which passages were low and necessitated a crawl, which were flooded, which had hidden pitfalls that would require careful stepping. (Wary of making the network *too* navigable, the elders left all entrances on the map unmarked.) Meanwhile, cataphiles over

the years had brought power drills and jackhammers underground to gouge out small passages from the walls:
chatières—"cat ways"—which would be vital gateways in
our trek.

Benoit, who wore only a small bag to hold a bottle of
water and an extra light, eyed our bulging packs. "How
long do you plan to stay?" he asked.

"We're hiking across the city," Steve said. "To the northern frontier."

Benoit stared at Steve for a moment, then laughed, evidently assuming it was a joke, before turning and heading
off into the dark.

WE TWISTED, SQUIRMED, AND CRAWLED, contorting our
bodies, as though performing an extended subterranean
gymnastics routine. We squeezed through long constrictive
passages, emerging in a tangle of limbs, like a newborn foal.

We climbed down into chambers the size of ballrooms, where our voices reverberated off the ceilings. The walls were slick with condensation and gave off steam: it was like moving through intricate foldings of brain tissue. We peered up manhole shafts seventy feet high, where it was too dark to see the top. Brown roots crept down from the ceiling like small, craggy chandeliers. The main tunnels were marked with Paris's signature blue ceramic signs, the names corresponding to the streets above. In a realm of palimpsests, the graffiti from spray cans of cataphiles obscured smoke streaks from torches of seventeenth-century quarry diggers, which obscured fossils of ancient sea creatures embedded in the limestone. Every few moments, we passed tunnels branching off on either side, a reminder of just how tangled our path would become.

Chris, Liz, and Jazz, the first-timers, walked as though in a dream. "I can't believe this place is real," whispered Jazz.

At one point, I shined my light upward to find a giant black crack in the ceiling. In the eighteenth century, there had been collapses: buildings and horse-drawn carriages and people walking in the street swallowed by the earth as the stonecutters below were buried in rubble. But today the tunnels were secure and we did not fear entombment—the catacombs were the least perilous leg of our journey.

LONG BEFORE NADAR BEGAN roaming the underside of Paris, he strove to photograph the world from unexplored perspectives: first, from the air. Together with his close

friend Jules Verne, Nadar founded the Société d'encouragement de la locomotion aérienne au moyen d'appareils plus
lourds que l'air (Society for the Encouragement of Aerial
Locomotion by Means of Heavier-than-Air Machines),
and launched spectacular hot-air balloon missions all over
Europe. In 1858, he boarded a balloon and rose above Paris,
where, from a height of 258 feet, he took the world's first
aerial photograph, a gently blurred silver-gray image of the
city. "We have had bird's-eye views seen by the mind's eye
imperfectly," he wrote of the aerial mission. "Now we will
have nothing less than the tracings of nature herself, re-
flected on the plate."

Now, for his next trick, Nadar would photograph the
city from below. It began with the arc lamp, which he had
assembled in his studio. It was a powerful, if unwieldy, contraption: upon activating a rig of fifty Bunsen batteries, an
electric current sparked two carbon rods, which sent out a
flare of white light. The lamp made it possible to create images without natural sunlight, a novel concept in the na-
scent medium of photography. In the evenings, he'd set off
the lamp on the sidewalk in front of his studio, drawing
crowds with the blaze. Nadar declared that he would use
his lamp to capture vistas with his camera that eluded all
other photographers. "The world underground," he wrote,
"offered an infinite field of activity no less interesting than
that of the top surface. We were going into it, to reveal the
mysteries of its deepest, most secret caverns." It was in the
ossuaries—called Les Catacombes, in imitation of the famous catacombs in Rome—that Nadar took his first subterranean photographs.

Seven hours or so into our journey, Steve guided us down a long passageway, and into a chamber of cobbled walls. Everyone unshouldered their packs and took a rest on the floor. Morale was high, despite wet feet and a considerable accumulation of mud. It was several moments before we identified the dry, copper-colored objects strewn over the ground at our feet.

Jazz picked one up and studied it, her dreadlocks shifting on top of her head. "It's a rib," she said, letting it fall from her hand.

Sure enough, we looked down to find that we were stepping over bones—a tibia, a femur, the crown of a skull, each desiccated and smooth, the color of parchment. We ducked around a corner and found ourselves at the foot of a giant tower: thousands of bones in a jumbled cascade, spilling down a chute from the surface. We were standing in an ossuary beneath the Cimetière du Montparnasse.

At the end of the eighteenth century, Paris was overflowing with corpses. The walls of the Cimetière des Saints-Innocents, the largest burial ground in the city, buckled, and corpses tumbled into the basements of neighboring homes. In order to prevent the spread of disease, the city decided to relocate its dead to the underground quarries, which had been expanding beneath their feet for decades. The chosen site was a three-acre stretch of empty passageways in the south, beneath a street called Tombe-Issoire. After a trio of priests had descended underground to officially consecrate the tunnels, the bones began their trip across the city: they were transported in wooden carts draped in black veils, then dumped into pits in the streets. In all, the remains of six million dead made the trip into the quarries. Laborers were sent underground, and given the interminable task of sorting the bones and arranging them in intricate friezes.

In December of 1861, Nadar—along with a team of assistants and two mining carts creaking with camera equipment—descended into the bone-lined corridors. The galleries had been briefly open to visitors in 1810, but were

closed soon after due to vandalism; when Nadar arrived, the tunnels had been shuttered for decades. In the "mole-hills," as he called them, Nadar encountered a crew of sub-terranean workers, toiling among the bones.

At the time, the process of taking a photograph even in the controlled environment of a studio was complicated; underground, in pitch-dark galleries, it was nearly impos-sible. The delays were maddening: the collodion formula would spill in the dark; the arc lamp would get stuck in tight corridors; the batteries would emit noxious fumes, which, in confined spaces, made everyone ill. As each shot required an eighteen-minute exposure, a full day of work brought few photographs; an assistant was heard to grumble, "We grow old down here." But Nadar worked furiously. As his model he cast a wooden mannequin, which he outfitted with a beard, hat, boots, municipal coveralls, and a pitchfork used to rake bones.

Nadar produced seventy-three photographs of the ossuaries, a uniquely quiet and surreal collection. One showed a freshly pitched pile of bones; others focused on the intricate bone friezes, or on mannequin-workers pushing bone-filled wagons through corridors. From the moment they went on display at the Société française de photographie, the images were a sensation. Critics wrote of Nadar as a mythical figure, traversing the cosmos of the city. An article in the *Journal des débats* called the photographer "Beelzebub," the lord of the underworld; another referred to him as a necromancer who had "electrified the mortal remains of past generations." A whole secret dimension of the city had been unveiled: "He and his assistants," wrote one journalist, "now in the bowels of the harmless earth, will enable people to become familiarized with scenes which but few have witnessed." Nadar became the toast of salons and cafés, as the city talked incessantly of the subterranean images.

But it wasn't only talk. The photographs awakened something in Parisians: having glimpsed the city's underside, they wanted to touch and smell the tunnels for themselves, to hear their own footsteps in the dark. Around the same time the photographs were first displayed, the catacombs were reopened, rapidly becoming one of the city's premier attractions. A few times a month, then more frequently, men in top hats and ladies in long dresses would wander through the ossuary in huddled groups, peering into hollow eye sockets of browned skulls, watching walls of stacked tibiae ripple in the candlelight. They shivered at the unearthly acoustics and the sensation of being under the damp earth; at the end of the tour, visitors would surreptitiously pluck skulls from the walls, stealing souvenirs from the underworld. So popular became the catacombs that in 1862, when Gustave Flaubert visited with the novelists Jules and Edmond de Goncourt, they grew cranky about the crowds. "One must put up with all those Parisian jokers," wrote the famously acerbic Goncourt brothers, "who go underground on veritable pleasure parties and amuse themselves by hurling insults into the mouth of Nothingness."

There was, too, a rush of unsanctioned visitors—proto-cataphiles—who pressed into the stretches beyond the route of the tour. Lovers arranged underground trysts; teenagers struck out on exploratory missions. Just as their cataphile descendants would do many years later, a group of Parisians held a clandestine concert in the catacombs. One hundred guests gathered on rue d'Enfer, having parked their carriages down the street so as not to arouse suspicion, then slipped down through the entrance. Sixty feet beneath the city, amidst candles burning atop human skulls, the guests sat before an orchestra of forty-five musicians. The evening's bill included "Funeral March" by Chopin and Saint-Saëns's *Danse Macabre.*

WE CAMPED AN HOUR NORTH, in a boxy chamber excavated at some point in the nineteenth century. From iron rings in the walls, everyone strung up hammocks, while Liz and I made spaghetti with tuna fish. We ate in silence, happy and exhausted. It was like camping on the moon: no sounds down here, nothing alive, just miles of darkness.

As we prepared for bed, Chris asked what time it was. Moe pointed out that we were in a place that had been perfectly dark and exactly 57 degrees Fahrenheit since it was created, a space untouched by all natural rhythms. "It's never o'clock," he said.

I awoke to find a woman standing in the doorway of our chamber. In one hand, she held an antique wrought-iron lantern with a hissing, naked flame that emitted honey-colored light. I watched her tiptoe into the center of the room, where she placed what appeared to be a small postcard on the floor.

"*Bonjour,*" I said, startling her.

Misty was in her forties and had been visiting the quarries since she was sixteen. Tonight she'd been wandering the tunnels alone—sans map, I noticed.

"Sometimes it's nice to just go down for a walk," she said in a lilting accent. Somehow her boots looked spotless, her gray blouse freshly dry-cleaned. As she walked from chamber to chamber in the quarries, Misty left little paintings wherever she went, like small messages to other cataphiles. The image she'd placed in our chamber showed two hands making a triangle.

———

IT WAS 1 A.M. WHEN we found our exit point from the catacombs: a *chatière*, barely wider than my shoulders. We were in a corner of the quarries seldom visited, where the ceiling was braced with centuries-old wooden bars installed long ago by the Inspection générale des carrières.

We'd now been underground for twenty-seven hours. I had dried mud in the folds of my ears, the rims of my nostrils.

"I feel like I'm turning into a troglodyte," said Liz, stretching out her legs in the tunnel.

"I keep picking stuff out of my hair that I don't recognize," said Jazz, studying a dreadlock. "I think I just found bone marrow."

Moe removed his sock, took out a small vial of iodine, and started brushing the bright orange liquid into his toenail cuticles. Steve blinked at him.

"You think I'm *not* going to sterilize my hangnail before the sewers?"

To get to the sewers, we first had to negotiate a stretch of utility tunnels that would take us under the Seine. If the catacombs were the city's cerebellum, the concrete tunnel we emerged into was a vein, a modest conduit linking more intricate organs. As we walked, it became clear just how close we were to the surface: filtering down from the street were the sounds of people chattering, high heels click-clacking, a dog barking. Through a vent in the wall, I caught an orange glow—the lights from an underground parking garage. I crouched and watched a woman with dark hair get into her car, back out, and drive away, and I felt like a ghost peering in on the city of the living.

We were unable to find a direct connection to the utility tunnel under the Seine, and so we had to go aboveground, but only for a moment. At the bottom of a manhole shaft with a ladder leading to the surface, we discussed the choreography of the exit in anxious whispers.

"I think I'm more nervous about getting caught than dying," whispered Moe.

"It's okay," Steve said. "If they put us in jail, we'll dig a tunnel out."

A twitch of concern registered in Chris's eyes.

We emerged near Saint-Sulpice, in front of a store selling luxury baby clothing. With no police in sight, we set off zigzagging through empty alleys, moving in the direction of the Seine. At the end of a deserted street, Steve crouched and popped a hatch, and we all slipped back underground. As I lowered myself down, I caught the eye of a late-shift busboy holding salt and pepper shakers, a look of bewilderment on his face.

The tunnel under the Seine was damp, with dismal, submarine acoustics. Even here, we found evidence of interlopers: traces of graffiti, an empty liter bottle of Kronenbourg beer. As we crossed beneath the river, I imagined a cross-section view of the city, showing each stratum, layered one on top of the other. Above us, the towering silhouette of Notre-Dame, the bridges, the river. Deep below, the tunnels of the metro, which would soon be teeming with commuters. And there we were in the middle layer, six tiny cones of light cutting through darkness.

———

THE DARK AND TWISTING SEWERS had, in the days before Nadar, been a source of irredeemable dread for Parisians. In Victor Hugo's *Les Misérables,* which was written during the two decades preceding the release of Nadar's photographs, the sewers embodied a kind of composite urban nightmare. The "intestine of the Leviathan," Hugo wrote, is "tortuous, with cracks and torn-up cobblestones and ruts and strange bends, rising and descending for no obvious reason, fetid, wild, ferocious, sunk in darkness, bearing scars on its paving stones and gashes on its walls, frightful."

In the 1850s, Georges-Eugène Haussmann, the famous city planner under Napoleon III, gave the sewers a full renovation. He gutted the city streets and laid four hundred miles of new pipes. Engineers installed each segment of pipe at a slope of three centimeters for every meter—gradual enough to allow for easy walking, steep enough to ensure a steady flow. In a series of tests, they discerned that an animal carcass ran the length of the city in eighteen days, while confetti made the same trip in six hours. But no amount of renovation could soften the public's aversion. Outside of sewer workers—the *égoutiers*—who spent their days scrubbing muck out of the pipes, no one willingly entered the sewers.

WE'D BEEN IN THE SEWERS maybe ninety seconds when the call came from Steve at the front of the line—"Rat!"

Gray and bandicoot-sized, it skittered up the stream of

wastewater at our feet. We all leaped up and straddled the sides of the pipe as it ran under us, sweeping its tail, kicking up a V-shaped wake.

Our route north would take us up the collector beneath boulevard de Sébastopol, a large, circular, brick-lined channel, flanked by two thick water pipes—one carrying potable water, the other nonpotable. It was into the collector that all of the smaller tributary pipes flowed. Down the center ran a recessed canal called a *cunette*—four feet across, misted over with vapor—which carried a flow of every conceivable form of matter rejected by the surface. A one-minute game of "I Spy": a syringe, a dead bird, a soggy metro ticket, a chopped credit card, a wine label, a condom, a coffee filter, many globs of toilet paper, as well as floating pieces of shit. "Sewer fresh," said Moe, *fresh* being urban explorer lingo for "human excrement."

Just as we were gearing up—Liz squirting hand sanitizer on everyone's hands, Moe powering up his gas detector—Steve called for our attention.

He'd received a text message from Ian, our weather sentinel.

RAIN FORECAST, POSSIBLE THUNDERSTORM,
LOOKS TO BE WET.

Steve went around the circle, looking at us one by one, but there was no hesitation. We were thirty-one hours deep: we'd come too far to give up.

"We just have to be vigilant," said Steve. As long as we kept a close eye on the level of water in the *cunette,* he said, and the water coming out of the tributary pipes, we'd be fine.

Steve was more familiar with the trappings of sewers than just about anyone on the planet, which was both comforting and unnerving, because he could narrate in sumptuous detail exactly what would happen to us in the event of a rainstorm. On the slimy wall of the collector, he traced a little graph with his finger, showing the exponential rate at which water would rise. "I've walked collectors in New York City, London, Moscow," he said. "But the flow in Paris is the burliest I've seen. Up to your shins, your knees, your waist, before you even realize it. The second we see water start to rise, we just bolt like crazy to the nearest ladder."

As we headed up the collector, no one spoke. I was practically tiptoeing: the catwalk was slippery and my shoes had almost zero traction. The air was jungle-thick, with burbles and gargles and belches rising up all around us, the sounds of Paris metabolizing. The stink was subtler than you'd imagine—the smell of a refrigerator that needed cleaning—but nonetheless a smell that would cling to you. At the dark junctions were Piranesian traps of slimy ducts and valves. Passing under one rig, maybe fifteen feet up, I saw small ribbons of tattered toilet paper—evidence that a flood had recently gushed down this very pipe.

At one point, a jet of water burst out of a tributary, sending a shock of echoes down the collector. We all froze, wide-eyed, preparing to dash to the nearest ladder.

"Nothing to worry about," said Steve. An early riser in one of the apartments above had flushed. "Everything down here is magnified," he reminded us. "Even a little spray is going to sound like Niagara."

NADAR BEGAN EXPLORING THE SEWERS shortly after his visits to the catacombs. Over the course of several weeks, he navigated the city's digestive system, his assistants lugging his equipment up and down the catwalks. Compared to the catacombs, the sewers presented far more logistical challenges. Here he was exposed to every vicissitude on the surface, every rain shower and dispersal of toilet water, making it more difficult to find the necessary eighteen minutes of uninterrupted stillness. Each time Nadar opened the shutter, he and his crew would pray that nothing would disrupt the shot. "At the moment when all precautions had been taken," Nadar later wrote, "all impediments removed or dealt with, the decisive moves being about to take place—all of a sudden, in the last seconds of the exposure, a mist arising from the waters would fog the plate—and what oaths were issued against the *belle dame* or *bon monsieur* above us, who without suspecting our presence, picked just that moment to renew their bath water."

Nadar's photographs of the sewers revealed the shadowy pipes with a romantic shimmer. Some featured the bearded mannequin, now sporting the coveralls of the *égoutier,* propped up in the poses of labor. Other images were abstract, focusing on geometric lines: a pipe splitting into two channels, or a current of sewage flowing in a ghostly blur. Due to the vapor in the pipes, each photograph bore a faint haze, as though cast behind a veil.

Journalists and reviewers again fawned over the images. A newspaper portrayed Nadar as a pioneer, fighting off peril and treachery in the city's long-maligned subterranean wilderness, making photographs despite being "half-asphyxiated by noxious gases from the electric battery in those suffocating vaults." The philosopher Walter Benjamin described them as "the first time that the lens is given the task of making discoveries."

All over Paris, people began popping sewer manholes. Late at night, they'd climb down, light a candle, and go for a stroll. An 1865 account of a midnight gambit in *La Vie Parisienne* imagined the sewers as a new promenade. "There are charming encounters to be had there. I encountered the pretty comtesse de T——, more or less alone, I also saw the marquise D——, and I rubbed elbows with Mlle N—— of

the Variétés theater." The day would come, the writer predicted, when the allure of sewers eclipsed that of the city's
verdant parklands. "When it becomes possible to tour the
sewers on horseback," he wrote, "the Bois de Boulogne will
undoubtedly be deserted."

During the Exposition Universelle of 1867, the city
opened the sewers to official tours, and visitors flocked
from all over Europe. Dignitaries and royals, diplomats and
ambassadors descended by an iron spiral staircase near the
Place de la Concorde and boarded a wagon otherwise used

by sewermen to clean the pipes. "A chariot with cushioned seats, its corners illuminated with oil lamps," recalled one visitor. Ladies in bonnets and high heels, holding lacy umbrellas, glided through the city's ejecta. The sewer workers played the part of gondoliers, pulling the boat down the canal. "Everyone knows," recorded a contemporary travel guide, "that no foreigner of distinction wants to leave the city without making this trip."

Meanwhile, Nadar embraced his role as the Hermes of Paris, the psychopomp, interlocutor between above and below. In the years after his photographs were released, he was known to give private tours of the sewers and quarries, leading tittering groups through the dark. In an essay accompanying his images, the photographer beckoned the masses to follow him into the depths. "Madame," he wrote, addressing one of his followers, "allow me to be your guide. Please take my arm and *suivons le monde*."

———

BEFORE THE FINAL STRETCH, we camped on the banks of an underground section of the Canal Saint-Martin: a broad, arched tunnel where green water flowed placidly, as filmy morning light filtered through the far end. It was about 8 A.M.—on the surface, brasseries would soon be opening, waiters arranging silverware on tables. We strung our hammocks to the railing along the canal, like alpinists in cliffside bivouacs. Steve volunteered to stay up and keep watch.

As I lay in my hammock, thinking of Nadar's photographs, I recalled a moment from the myth of Phaëthon, the young man who convinces his father, Helios, to let him drive the flaming sun chariot across the sky. Soon after lifting off, the boy loses control of the horses' reins: the vessel swerves toward the earth, the heat dries up rivers, creates

deserts, sets fire to mountaintops, until, finally, Phaëthon flies so low that the chariot burns a hole clear through the earth's surface, allowing light to pour into the underworld. People scramble to the edge of the hole, where they find that they can see, for the first time, straight down into the sweep of Hades, from the fire-ringed lakes and the gloomy Asphodel Meadows to the endless black of Tartarus. They even see King Hades and Queen Persephone, seated on their thrones, blinking up at them. The people are terror-struck by this infernal landscape, which they have so long dreaded; and yet, they don't retreat from the edge of the hole. They continue to peer into the gloom, unable to look away.

We'd been asleep maybe two and half hours when Steve spotted a tour boat gliding down the canal. Before the boat captain spotted us and called the police, Steve shook everyone awake and we slipped back into the dark.

Our final stretch was the collector under avenue Jean Jaurès—a long, squared-off corridor, broad and plain, with a current of sewage the width of a one-lane road roaring down the middle. According to Steve, we were walking the main line: almost all of the wastewater in Paris was flowing at our feet.

Now at the thirty-eight-hour mark, we could all feel our destination near. We might have felt triumph, or relief, or a sense of accomplishment, but we were dragging our feet, eyes bloodshot, all of us haggard and a touch dazed, I suspect, from whatever subterranean miasma we'd inhaled in the preceding hours and miles.

"I say we push to the north of France," Steve said.

Down the slick catwalk, I could feel my eyelids starting to droop: I kept as close to the wall as possible and focused on putting one foot in front of the other. Every few hundred feet, we passed a small tributary pipe marked with the sign of the street above it. Moe, walking in front with the map, called out each street name, and counted off the distance until our destination.

"Five hundred meters!"

With every step, the gush of the canal grew stronger, the sewage splashed over the edges of the catwalk, before it was lapping at the tops of our shoes—the underground was pushing us out.

WE CAME TO THE surface under a bright midday sun, just beyond the city limits: six of us ascending a ladder to emerge through a manhole at the foot of a Turkish restaurant. Our faces were smudged, our hair matted with muck and slime, our clothes soggy and fetid. As we emerged, pedestrians on the sidewalk halted and jumped back, a waiter at the restaurant dropped a fork and knife. An old woman in a pink sweater leaned on her walker and stared down at us, eyes wide, her mouth in a perfect O. And just for a moment—before Steve pulled the manhole cover back into place, before we all stumbled into a nearby park and slashed open a celebratory bottle of champagne—everyone on the street was leaning forward to peer down into the open manhole.

THE INTRATERRESTRIALS

The stone's alive with what's invisible.

—SEAMUS HEANEY,

"SEEING THINGS"

In April of 1818, a man from Ohio named John Cleves Symmes declared his intent to lead a voyage to the interior of the earth. Symmes, a thirty-eight-year-old retired Army captain who ran a trade outpost in the frontier town of St. Louis, sent a formal mission statement to more than five hundred dignitaries, including congressmen, scientists, newspaper editors, professors, museum directors, and several reigning European princes. "TO ALL THE WORLD," he began. "I declare the earth is hollow and habitable within; containing a number of solid concentrick [*sic*] spheres, one within the other." The inner world, he proposed, was populated by enigmatic, unknown life-forms, perhaps undiscovered races of human, and that it could be reached by enormous circular openings at the North and South Poles. "I pledge my life in support of this truth, and am ready to explore the hollow, if the world will support and aid me in the undertaking." He concluded with a call to arms:

"I ask one hundred brave companions, well equipped, to start from Siberia in the fall season, with reindeer and sleighs, on the ice of the frozen sea; and I engage we find a warm and rich land, stocked with thrifty vegetables and animals, if not men."

The captain's declaration was met with silence: no wealthy prince lent his support, nary a "brave companion" stepped forward. Undeterred, Symmes embarked on a lecture tour to drum up support for his "new theory of the earth." He traveled the frontier in an old dusty carriage, moving from town to town. Outside of saloons and meeting halls, he'd set up an array of props to help illustrate his theory: metal filings and magnets, bowls of spinning sand, a wooden globe with openings at the top and bottom. For hours, he'd regale audiences with tales of the undiscovered lands beneath their feet.

For a fleeting moment, Symmes was celebrated. A small, twitchy man, he proved to be a dynamic performer. Audiences enjoyed the captain's vision of the inner earth: a whole new frontier waiting to be explored and incorporated into the growing American union. "The Newton of the West," they called him. Word traveled and Symmes's tales of the hidden earth were covered in newspapers and magazines. Which is about when the public took a closer look at the science behind his theory, and realized just how ludicrous it was.

Based on the fact that Saturn possessed concentric rings, Symmes had concluded that concentricity was a universal design in nature, and therefore "all planets and globes must be hollow," composed of nested spheres. Symmes was dismissed as a charlatan. "The theory was overwhelmed with ridicule as the product of a distempered imagination or partial insanity," wrote one historian. "It was for many years a fruitful source of jest."

Even as his theory was jeered, the captain continued lecturing, drafting petitions to Congress, and seeking financial support for his expedition. In 1823, he convinced the Russian chancellor, one Count Romanoff, to fund an inner-earth mission, but at the last moment, the count got cold feet and backed out. In 1829, during a lecture tour through Canada and New England, the forty-eight-year-old captain fell ill; he died in the back of his carriage, rolling west. At the end, he was roundly considered a loon: a man who'd wasted his life chasing fairy tales of underground lands and intraterrestrial beings.

In the years after Symmes's death, however, the Western world became fixated on stories of exotic subterranean life-forms. The captain's theory found expression among a variety of novelists and artists. Edgar Allan Poe championed Symmes's theory, using it as a motif in a short story, "MS. Found in a Bottle," as well as a novel, *The Narrative of Arthur Gordon Pym of Nantucket*, about a sailor's journey into the inner world. Jules Verne, meanwhile, adapted the hollow earth theory in *Journey to the Center of the Earth*, in

which Professor Lidenbrock descends through an Icelan-
dic volcano into a hidden world inhabited by ancient sau-
rian beasts. The novelist H. G. Wells; the creator of Tarzan,
Edgar Rice Burroughs; the author of *The Wonderful Wizard
of Oz,* L. Frank Baum; and countless others told stories of
inner worlds. In the last decade of the nineteenth century,
more than a hundred novels about underground life-forms
were published in the United States alone.

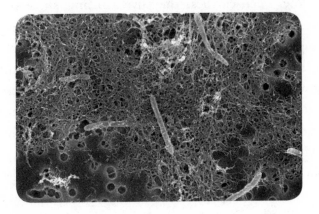

Captain Symmes was one of my first underground heroes. For a while, I kept a cutout of his face pinned above
my desk. Rather than a failed scientist, I took the captain
as a poet, a surrealist who told stories that seemed out-
landish on their face, and yet contained mysterious seams
of poignancy. I was curious about the way Symmes's vision of intraterrestrial life had wormed its way into the
collective imagination: it almost felt like he'd reached
down and touched an ancient truth, pressing on deep,
shared memory.

It was Captain Symmes I thought of one summer when

I came across a story about a team of biologists who'd climbed to the bottom of an oil borehole in the desert, almost a mile beneath the surface, and found something curious: living creatures. They were bizarre, wriggling, single-celled bacteria, and they were living far deeper underground than should have been possible. As it turned out, all over the world, I learned, biologists were finding similar microbial creatures in caves, abandoned mines, and other deep cavities. They lived in impossible conditions, where all other creatures would wither—absolute darkness, scalding hot temperatures, intense atmospheric pressure, with little oxygen, and even less food. So divergent were they from known life, the subterranean microbes might as well have come from a distant planet; indeed, NASA had begun to study them as possible analogs for life on Mars. It turned out that they were everywhere, even down *inside* the earth's crust, in groundwater flowing through microscopic passageways of porous rock. And they were old: some had been living, cut off from the surface world, for millions of years. If only dear Captain Symmes had lived to see the teeming super-tribe of mysterious, ancient organisms thriving in the inner earth.

What amazed me most of all—and what would've really delighted the captain—was that some microbiologists believed these deep-dwelling creatures were related to the planet's very first life-forms, that life had originated underground. Rather than the warm pool of water on the surface that scientists had long envisioned as the cradle of life, these researchers were suggesting that earthly life took root in the underworld: that hyenas and hedgehogs and horseshoe crabs and hippopotamuses and humans were descended from microbes that evolved deep in the crust and long ago emerged on the surface.

I was taken with the idea that in some buried part of us, we might hold a phantom imprint of our ancestral subterranean origins. And so I set out to meet a team of microbiologists working on an experiment for the NASA Astrobiology Institute called Life Underground. When I tracked them down, they were in South Dakota, hunting for intraterrestrials in the depths of an abandoned gold mine called Homestake, a mile beneath the surface—far deeper inside the earth than I'd ever been.

ON A SPRING AFTERNOON, under a broad dome of blue sky, I followed a sidewinding road into the Black Hills. I moved through a landscape of golden lakes fringed with ponderosa pine, down into meadows rutted with boulders, across prairie lands thronged with buffalo.

The Black Hills are a thumbprint-shaped mountain range that covers some 4,500 square miles, mostly in western South Dakota, with the northwestern edge spilling into Wyoming. They encompass some of the oldest stone in North America, with granite and sandstone that cracked up out of the surrounding flats some seventy million years ago. Covered in dark pine, spruce, and fir trees, the range holds a dark silhouette against the pale sagebrush of the Great Plains. Perhaps due to the way they rise up like a giant spiky beast, or the way lightning spasms over their peaks, the Black Hills have long inspired spiritual awe. One nineteenth-century visitor referred to them as the "abode of the geneii, or thunder spirits, who fabricate storms and tempests."

For Native American tribes of the Great Plains, who have been roaming the territory since at least thirteen thousand years ago, the Black Hills have always been a sacred destination. It was a place to hunt buffalo and antelope, to gather medicinal plants, and to harvest pine trees. Native people would venture into hidden canyons in the hills to carve petroglyphs into stone walls, and to embark on vision quests to communicate with the spirit world. The tribe most intimately connected to the hills is the Lakota, who claim the territory as their ancestral homeland, the place from which their ancestors emerged. They call it Wamaka Og'naka Icante, which means "Heart of Everything That Is."

Before leaving my apartment in New York, I'd thrown into my bag an old text on the beliefs and customs of the Lakota. The book—based on the notes of James Walker, a physician who worked on the Pine Ridge Reservation at the end of the nineteenth century—had been an afterthought: a distraction from the stack of articles on subsurface microbiology I'd planned to read on the flight. But I started reading it at takeoff and couldn't put it down. The Lakota, I learned, were a curiously subterranean-fixated culture. An old map of sacred destinations in the Black Hills, drawn by a Lakota artist named Amos Bad Heart Bull, showed a constellation of underground spaces. In the southwest part of the hills, for example, the tribe gravitated to a cluster of sacred hot springs, and conducted rituals around a sinkhole into which their ancestors drove buffalo. In particular, they gathered to perform ceremonies at the mouths of caves, especially at Washu Niya, or "Breathing Cave," known to white people as Wind Cave, which is one of the largest and most dazzlingly intricate caves in the world. Each of these openings was seen as a conduit to the otherworld, a portal between earthly and spiritual realms. As I weaved my way into the northeastern part of the hills to meet with Life Underground, my arrival felt unexpectedly backlit by the Lakota beliefs.

THE LIFE UNDERGROUND TEAM began to plumb the hidden sphere of life in 2013. Under the leadership of University of Southern California biologist Jan Amend, they assembled sixty scientists from five other institutions—the California Institute of Technology, the Jet Propulsion Laboratory, Rensselaer Polytechnic Institute, Northwestern University, and the Desert Research Institute—and began probing the earth. They went deep into boreholes and mineshafts all over the world, as well as natural springs, even beneath the floor of the ocean. In each place, they collected samples, which they brought back to examine in the laboratory. The ultimate objective was to search for microbial life on Mars, which they believed was most likely to dwell in the subsurface, where it would be protected from harsh conditions aboveground. But before anyone began rooting around for underground life on the Red Planet, they wanted to become better acquainted with the subterranean-dwellers here on earth, to understand how these improbable creatures had carved out a living in the underworld.

IN THE PARKING LOT of a Motel 6, just beyond the strip of old-timey casinos in the town of Deadwood, I climbed into a Jeep with two members of the Life Underground team. In the driver's seat was Brittany Kruger, a geochemist out of the Desert Research Institute in Las Vegas. She was in her early thirties, blue-eyed, with a long, blond ponytail and chiseled rock-climber arms. As a field biologist, she told me, she'd spent "an entire life going places and getting filthy." Next to her was Caitlin Casar, a geobiologist from Northwestern University: tall, thin, and laid-back, with short brown hair and gauges in her ears. The fourth member of our team was Tom Regan, who would be our guide into the depths of the mine.

We drove into Lead (rhymes with *seed*), which would be a small western town like any other—rows of small houses and low-slung municipal buildings—but for the enormous gaping maw at its center. Indeed, it was more hole than town. The Homestake Gold Mine Open Cut is the only section that is visible from the surface. At about a half mile wide and 1,250 feet deep, it is impossible to see the bottom from any point on its perimeter. (For five dollars, at the Homestake Gold Mine Welcome Center, you can hit a golf ball over the lip.)

Homestake originated with a sordid and shameless land grab by the U.S. government. In 1868, the government signed a treaty granting unequivocal ownership of the Black Hills to the Lakota, declaring that white people could not enter the territory without Lakota permission. But when rumors of gold drifted out of the hills six years later, the treaty was promptly scrapped, and people rushed in and started digging. Homestake, opened in 1877 by the tycoon George Hearst, was their largest excavation. Over the next century and a half, it grew to be the most productive mine in the western hemisphere: at eight thousand feet deep, with 370 miles of tunnels, it was a kind of industrial Grand Canyon. In 2001, the mine stopped being profitable; the pumps on the deep levels were shut off and the pits began slowly filling with water.

Homestake sat dormant until 2012, when the owners of the mine reopened the space as a science laboratory: Sanford Underground Research Facility, or SURF. It was ideal for physicists to conduct experiments deep under the surface, where the mass of rock became a natural filter of cosmic radiation. The day I visited, there were fourteen active experiments at SURF, most of which were set in renovated portions of the mine, with fluorescent lights, bright tile floors, and graduate students on laptops. And then there were the remote parts, the dark wilderness of the mine, down on the 4,850-foot-deep level, where the rock remained untouched and raw, where hot steam seeped from the walls—this was where we were headed.

DOWN IN A CONCRETE-SHELLED CORRIDOR, the Life Undergrounders and I waited for the elevator—known as "the Cage"—that would take us into the depths, first to the eight-hundred-foot level, then down to the bottom. Employees of the facility passed by: former miners, burly men who now worked maintenance in the tunnels, and physicists—skinny guys who spent their days in the laboratories. We all adjusted our equipment—bulky blue coveralls, helmets, headlamps, safety goggles, steel-toed rubber boots, and a self-contained respirator, which was a kind of external lung packed into a grenade-sized canister, to be activated only in the case of a fire or gas leak.

"When we get down, it might feel a little raw," said Tom Regan. "But as long as you stay calm, everything will be fine." The SURF safety specialist, Tom was in his late sixties, a short, bespectacled man, a Vietnam vet, a deacon at a church up the road in Spearfish. In these first moments, Tom left little impression on me. He spoke largely in safety-related acronyms, enumerating protocol for various potential accidents. Not that he was unpleasant, but he struck me as weary, stilted, a little detached. In any case, I didn't pay him much mind because I was feeling anxious about our descent.

I'd never been more than a few hundred feet under-ground, and now we were going much, much deeper. From time to time, I'd been told, visitors to the raw parts of the 4,850-foot-level had lost their nerve: the absolute darkness, or the feeling of enclosure, or the hard fact of a mile of granite overhead, had triggered a psychic crack-up, and the person had to be whisked back up to the surface. It made me think of an old story about a group of early cave explor-ers in England who lowered a companion by rope into the pitch-darkness of a vertical cave. As the man passed into the dark zone, they heard a terrible scream and quickly hauled him up: the man's eyes, as the story goes, were roll-ing in his head and all of his hair had turned gray. I fiddled with the clip on my emergency respirator, thinking of just how physiologically unfit we are for dark worlds, how we are aliens underground.

The Cage rattled into view, the door opened, and we all stepped into a large steel box with metal grilled walls. The operator—the Cager—was a rhinoceros-sized man in cov-eralls, with soot-smudged cheeks. He shook Tom's hand, then grinned at Brittany, Caitlin, and me. "What are we looking for underground today?" he hollered over the roar of the engine. "Or are we just going for a little walk?"

Brittany bellowed, "Microbes!"

The Cager burst out laughing and shook his head.

He pulled a lever, the doors clanged shut, and he called out, "Going down!" The Cage rumbled, lurched, and started to descend into pitch-darkness. I looked down at the floor. As my headlamp illuminated the open grillwork, I became acutely aware of the mile of empty space falling away at our feet. The rocky walls of the shaft began to slide past, slowly at first, then faster, as we rocketed down into the earth.

HUMANS HAVE LONG BEEN captivated by the possibility of subterranean creatures leading a hidden, parallel life in the underworld. In his fifth-century B.C. *Histories,* the ancient Greek historian Herodotus described a race of people who lived in the darkness of caves in Ethiopia. The Troglodytes—from the Greek *troglo-* (hole) and *-dyte* (to get or go into)—were recorded as nocturnal albino pygmies who ate lizards and "screeched" when exposed to sunlight. Herodotus reported many apocrypha in his *Histories*—tales of dog-sized ants who dig for gold in India, for example—but the Troglodytes persisted. Despite a complete lack of eyewitness accounts, subterranean-dwelling humans were referenced again and again, from Strabo and Pliny the Elder, all the way up to Carl Linnaeus, the eighteenth-century Swedish botanist responsible for establishing the standard Latin taxonomic classification of the natural world. Linnaeus declared that there were two separate species of human: one on the surface, one underground. *Homo diurnus,* or "day man," made his living off of sunlight and oxygen aboveground; in the depths of caves, *Homo nocturnus,* or "night man," lived in the dark and hunted nocturnally. The reality of darkness-loving, subterranean-dwelling humans gradually faded, but the possibility of a hidden counterpart had clearly touched a nerve, as though we were unconsciously searching for our inversion, our shadow-self.

The first confirmed discovery of subterranean-dwelling life came in 1689, when a nobleman from Trieste, Baron Johann Weikhard von Valvasor, published a historiography of Slovenia. In his description of the cave-riddled region called Karst, Valvasor described a serpentlike animal, about a foot long, that was flushed out of cave mouths during heavy rainstorms. The creature was familiar to locals, who believed it was the undeveloped spawn of underground-dwelling dragons. Valvasor called it the olm: an aquatic salamander, which lived full-time underground. In *The Origin of Species,* Charles Darwin cited the olm as an exemplar of his theory of adaptive evolution: once part of a surface-dwelling population, it began spending more time in underground environments, perhaps seeking shelter from predators, and gradually, over the course of millions of years, the physical traits beneficial to subterranean life were passed on. In the food-scarce underground environment, it had developed a remarkably efficient metabolism, enabling it to go a full year without eating. Meanwhile, in perpetual darkness, where it needed no protection against UV rays, the olm lost its pigment, its skin turning a cadaverish ivory. Even its eyes were rendered obsolete, completely obscured beneath a cover of skin.

Before long, biologists had identified multiple classes of cave-dwelling creatures. "Shade animals," who lived at the mouths of caves; "twilight animals," who lived within the reach of diffuse light; and, finally, "dark zone animals," or *troglobites,* like the olm, who were so seamlessly adapted to subterranean life, they couldn't survive on the surface. Cave expeditions revealed a dreamlike bestiary of troglobites: albino catfish, pearlescent spiders, blind beetles, see-through crabs, and eyeless insects. The troglobites, it was assumed, were the sole occupants of the underground: surely no other creature could survive in the dark zone.

The doors on the underworld kingdom blew open in 1994, when a young biologist from New Mexico named Penny Boston climbed down to the very bottom of Lechuguilla Cave, two thousand feet underground. It was an environment, she said, "as close as you can get to traveling to another planet without actually leaving earth"—far too remote to support even the hardiest troglobite, or any other living creature. But at one point, Boston was scrutinizing a furry, brown geological growth on the ceiling of a cave passage, when a drip of water plopped directly into her eye. Boston was amazed to find that her eye puffed up and swelled shut. It could only have meant one thing: she had been infected by bacteria, by tiny microorganisms living in the cave's depths, far deeper underground than anyone imagined possible.

Now researchers wondered about the *rest* of the underground world. That is, the unseen territory beyond caves, down in the rocky crust, which was considered solid by human scale, but was in fact sponged with tiny pores and fissures that coursed with groundwater. Even as the scientific community scoffed at the suggestion of life in the crust, where it was too dark, too hot, too high-pressure, and too empty of food, a few microbiologists went looking. They lowered themselves into gas boreholes, oil wells, and other man-made cavities, even drilled their own pits, in order to extract water samples from the depths. Sure enough, everywhere they looked, they found vigorous communities of bacteria. A thousand feet down, then a mile, then two miles deep, in treacherous, mephitic places, where pressure was four hundred times what it was at the surface, and temperatures approached 200 degrees Fahrenheit.

As more discoveries rolled in, biologists reckoned with the staggering scale and diversity of underworld life, which called for a drastic shift in perspective. Just as Copernicus pulled the earth from the center of the universe, and Darwin yanked humans from the center of history's arc, these discoveries suggested that surface-dwelling life was perhaps a minority on earth. The collective biomass of inner-earth life appeared to be almost equal to, perhaps greater than, that of surface life. If you placed all of subsurface microbial life on one side of a scale, and on the other side, you placed every surface-dwelling plant and animal, the scale would wobble. "We shake our heads incredulously," wrote the soil ecologist David Wolfe in 2001, "at the possibility of another living world, a hidden subterranean biosphere, more immense than the grand scale of life aboveground."

The "intraterrestrials" contradicted everything biologists had once held to be true about the characteristics of life. They did not breathe oxygen, did not rely on sunlight or photosynthesis for energy, did not consume carbon-based food. They subsisted on what biologists called a "dark food chain": eating rocks, or metabolizing chemical energy and radioactivity coming out of the earth's crust. They were our evolutionary alter-egos, a real-life version of a mysterious tribe from a hollow earth novel. Indeed, when a team discovered a species of bacteria two miles deep in a mine in South Africa, they named it *Desulforudis audaxviator*, or "bold traveler." It was an allusion to *Journey to the Center of the Earth*, in which Professor Lidenbrock's inner-world adventure is launched after he decodes a runic message about a hidden gateway to the planet's interior: *Descende, audax viator, et terrestre centrum attinges* (Descend, bold traveller, and you will attain the center of the Earth).

WE BID THE CAGER FAREWELL at the eight-hundred-foot level and disembarked into a narrow, rocky tunnel. Water poured down from the ceiling, drumming on the tops of our helmets. Behind us, the Cage rattled out of view and everything fell quiet. I stooped beneath the low, craggy ceiling, stepping through a slurry of mud up to the shins of my boots.

Brittany took the lead, followed by Caitlin and me,

while Tom brought up the rear. The air smelled sulfurous, and our headlamp beams caught fog in the dark. We were inside schist, a gray rock veined with yellow and orange. Mining carts full of rock had once rattled up and down this path, and now we passed signs reading DANGER and SEC- ONDARY ESCAPE ROUTE. As we walked, I thought of the eight hundred feet of solid rock overhead, and felt my heart slightly quicken; I wondered how my body would react once we descended another four thousand feet, to the bot- tom of the mine.

"I love it down here," Tom said. "When I'm under- ground, I'm just in my place." I did a double-take over my shoulder. The man who'd been so quiet and retiring on the surface, who'd spoken monotonously in acronyms and safety platitudes, now had a large, glowing smile on his face. As we continued walking, he became loose, warm, expansive—bubbly, even. It was as though he'd been hold- ing his breath and was just now exhaling.

He told of how he'd grown up in the hills, had gone away to Vietnam, then come back to work at the mine, how he'd worked in every possible role, from cager to test driller, how he'd found solace underground.

"I know the underground better than I know the streets in town," he said, pausing to touch a small protuberance in the rocky wall. "When they give me time off, and I don't go underground for a while, I start to get antsy. My wife and I will drive around the Black Hills and visit caves. If you haven't been to Wind Cave, it's just the most beautiful place you could imagine."

We heard a soft rumble from a remote precinct of the

mine, like a stampede of animals in the distance. "You hear that?" Tom said, softly. "You can just feel the way the mine shifts and settles. Like it's alive down here, and the tunnels are breathing."

We arrived at our first sample site, a metal pipe in the rock wall, maybe two inches wide, with water pouring out in a steady stream. The "seep," as it was called, had originally been dug to test for gold in the early 1900s, using a drill bit studded with industrial diamonds. There were a few dozen seeps in the mine; this one, Tom said, had been flowing uninterrupted for more than a century.

Brittany and Caitlin dropped their backpacks in the mud and got to work, adjusting their headlamps and goggles, which were already smeared with mud. They snapped on purple latex gloves and started unloading vials, graduated cylinders, and sensors used to measure the chemical makeup of the water, the temperature, the pH. To the end of the pipe, they rigged up a multi-tiered syringe, which enabled them to collect water samples uncontaminated by the air in the tunnel.

"A seep is like a tiny window into the subsurface that allows us to see what's living down here," said Brittany over her shoulder. "Water moves through the crust in big cycles, sometimes taking thousands of years to travel from one place to the next. Our idea is that the water from the seeps is coming up from a sequestered body of water down in the crust. Which would mean that any organisms we see are coming up from deep, deep down."

It would be weeks before the lab results would tell them exactly what was living in this water. But based on previous samples, in the various seeps in the mine, they expected to find species in the *Desulforudis* family: cousins, basically, of the bold traveler from South Africa.

When Caitlin and Brittany finished collecting samples, they hung a sign over the seep—NASA ASTROBIOLOGY IN-STITUTE, PLEASE DO NOT DISTURB. As they packed their equipment back into their bags, the sound of water eddied along the rock walls. I crouched and held my hand under the seep, feeling the water on my fingers, wondering at the depths it was bubbling up from. It was a moment before I realized that Tom was standing over me.

"I have some Lakota friends in Spearfish, folks from my congregation," he said. "They say that the water in the Black Hills is sacred, that the underground is connected to their ancestors."

What I ought to do, Tom told me, was hear the Lakota

creation story—the tribe's account of their origins. "I know the story some, but it's really not mine to tell," he said. "You need to talk to someone from the tribe. You should go to Wind Cave."

THE STORY OF THE ORIGIN of life—at least the way Western scientists have long told it—began about four billion years ago. A "primordial soup" of simple biochemical elements was sparked with energy, which caused them to combine into simple organic compounds: these became amino acids, which clustered into DNA and proteins, which eventually evolved into single-celled bacteria, which were the ancestors of all of life. Going back to Darwin, researchers believed that the stage for these primeval events was a shallow body of water: a tide pool, or a pond, or perhaps the calm surface waters of the ocean.

In 1992, a radical new theory emerged. It came from a retired researcher at Cornell University named Thomas Gold, who was an astrophysicist by training, but had a knack for exploring other scientific disciplines and presenting bold, iconoclastic theories, which frequently turned out to be correct. After years of tracking our encounters with intraterrestrials, he wrote a book called *The Deep Hot Biosphere,* where he argued compellingly for the abundance of underworld life. But then Gold went a step further: he proposed that life *began* underground.

Four billion years ago, Gold pointed out, the surface of the earth was a war zone. It was being inundated with lava from volcanic eruptions, baked under intense UV rays, and pummeled with a barrage of asteroids. It was extremely improbable, Gold argued, that the original delicate reactions of life—the first "gentle contacts," as he wrote—could have occurred amidst such tumult. The subsurface, on the other hand, was stable: no weather, no harsh light, no violent seismic activity. It was far more likely that our "Garden of Eden" was deep underground, where the earliest single-cellular microbes lived off chemical energy that rose up from the depths.

By Gold's model, the intraterrestrials—the oxygen-allergic, heat-loving, rock-eating darkness-lovers—were not a mysterious offshoot of us surface-dwellers. They came first: *we* were the offshoot of *them*. With that, Gold submitted an entirely new scene in our creation story: after millions of years gestating down inside the warm earth, a cluster of archaic microbes split off from the rest of the underworld inhabitants and slowly migrated upward, until they emerged into the light, where they gradually began to propagate aboveground. "Microbe pioneers," Gold wrote, "invaded the surface from below."

Over the last twenty-five years, evidence has mounted in support of Gold's theory. Microbiologists have encountered life deeper and deeper inside the earth, and in increasingly ancient pockets of water, perhaps as old as a billion years. Meanwhile, they are finding commonalities in the DNA of intraterrestrials, even in species living on opposite sides of the planet, as in the case of the *Desulforudis* dwelling in the depths of Homestake, which may suggest a possible shared ancestry. "It's still hard to say anything for absolute certain about life in subsurface," Caitlin told me.

"We've seen such a tiny pocket of what lives down there." But with each year, more and more microbiologists are accepting the possibility that life rose out of the earth.

It is a story they recognize, because it is a story we all know, and have always known—one of humankind's very oldest stories. Even as the underground world is the realm of death, it has always been a womb: a fertile and generative place, from which life emerges. Here lies the ultimate enchantment of the earth, where plants take root as seeds in the soil before sprouting onto the surface, just as all of us grow in the cave of our mother's womb, before emerging through a dark tunnel into daylight. In antiquity, cultures in every part of the world told subterranean creation stories—anthropologists call them "emergence myths"—where the primordial ancestors gestated down in the earth, before rising up into open space. We find them everywhere from Aboriginal Australia to the Andaman Islands of India to the folk traditions of Eastern Europe, but they were especially prevalent in the ancient Americas. According to the Hopi and Zuñi in the Southwest, for example, the first humans originated underground, in the deepest womb-world, in a kind of larval state, and as they ascended through subsequent womb-worlds, they became gradually more human, until they emerged on the surface, through the birth canal of their mother. In the tribes of central Mexico, meanwhile, the first humans emerged from the musky depths of a cave called Chicomoztoc, which translates to "the Cave of the Seven Niches." In faded Mesoamerican codices, we still see this cave-womb, lined with seven chambers, each containing tiny humans in fetal position, with a trail of footprints leading out of the cave. It is a tale that Bachelard would place "at the origin of all beliefs." When archaeologists squirm down into caves in France, they find thirty-thousand-year-old carvings of vulvae, marking the muddy depths as our place of origin.

NOW WE PLUNGED TO the very bottom of the mine—a mile underground. Next to Tom in the Cage, I watched the stone walls rush past in a blur—each minute, down another five hundred feet—and I felt my body reacting to the environment: the pressure like a yoke on my shoulders, the air thickening, causing sweat to prickle on my neck. Perhaps in these depths, with a mile of rock hanging over my head, my nerves would crack.

But that was not what happened at all. At the bottom of the mine, we stooped down a narrow, low-ceilinged gallery, where the rock walls were braced with rusted metal strips, until we reached the seep. As I crouched alongside Caitlin and Brittany, I watched water spill out of the rock, collecting in a large puddle at our feet, and I considered that the water was coursing with underworld creatures, that I was watching an upwelling of earth's archaic life-forms. The tunnel was hot, with steam rising out of the walls, but rather than oppressive, it struck me as generative heat, garden heat. However unnatural the environment may have been, however physiologically hostile and removed from the sphere of ordinary experience, this tunnel was a place of genesis. Watching the seep alongside us was Tom, who had spent five decades walking up and down these galleries, who felt more comfortable underground than above, who felt at home down in the earth. He was whistling softly to himself, holding an air of serenity, as though the stony walls were embracing him.

WITH ALL OF THE SAMPLES collected and our gear gathered, we made our way back through the tunnels and boarded the Cage. As we rumbled and rattled back up through the rocky shaft, no one said much, each of us inside our own thoughts. The moment we reached the surface, and stepped into the evening light, I felt limp with exhaustion. Caitlin, Brittany, and I zipped out of our muddy coveralls, stowed our goggles and helmets in the SURF locker room. As we loaded packs and equipment into the trunk of the Life Underground Jeep, Tom came to see us off.

Back on the surface, he seemed to have lost some of his glow, his skin becoming gray and wan. We shook hands and I thanked him for guiding us underground. "Do you need directions?" he asked.

I asked what he meant.

"To Wind Cave," he said. "It's not far. Just get on the main road from here, drive south down through the hills, and follow the signs."

THE FOLLOWING MORNING, I threaded between stony out-croppings, past the peak where the legendary Lakota sha-man Black Elk underwent his vision quest; past the stone monument to Crazy Horse, where a sculptor had chiseled the face of the Lakota leader from the side of a mountain, which, when it was finished, would be ten times the size of Mount Rushmore; past grazing buffalo and prairie dogs and a coyote slinking through the high grass; until, finally, I arrived at the heart of a golden meadow, where I met a woman named Sina Bear Eagle.

Sina was Oglala Lakota, the great-great-great-great-granddaughter of one of the tribe's renowned leaders, Chief No Flesh, whose photograph I'd tracked down in a mu-seum archive. She had grown up on the Pine Ridge Reser-vation, on the edges of the Black Hills. On her forearm, she had a Bob Dylan tattoo, and the bottom fringe of her shoulder-length hair was dyed turquoise. Sina welcomed me and led me down a winding path to the mouth of Washu Niya, or "Wind Cave."

The story, Sina said as we walked, was that the cave was discovered by the Bingham brothers—white men—in 1881. "But *discovery* is the wrong word," she said, her voice as gentle as it was commanding. "Lakota people knew about the cave long, long before."

Sina was about thirty years old and a woman of rising prominence in the Lakota community. She was a graduate student in linguistic anthropology at UCLA, where she was studying the Lakota language; when she finished, she planned to return to the reservation to teach kids their ancestral tongue. During the summers, she worked here at Wind Cave as a guide, teaching visitors about Lakota culture and the tribe's relationship to the cave.

An entrance to the cave had been opened for tourists, fitted with concrete steps leading down into the dark, but Sina and I sat just to the side, by the original cave mouth. It was a small opening, perfectly dark, maybe two feet in diameter. On days when the barometric pressure on the surface was lower than in the cave, Sina explained, you could feel air blowing out from underground. To demonstrate, she held a ribbon over the mouth, as the tail snapped outward.

"The cave has great *wakan*," she said, using the Lakota word for "holiness." She pointed out a bush next to the cave mouth, where the branches hung with small tobacco pouches, which Lakota people had left as offerings. On her very first visit to the cave, Sina told me, she, too, had left an offering. It was on a school trip, when she was twelve years old: as she descended underground, she knew immediately that she loved the cave. "I knew I wanted to come back here again and again," she said, "so I took a strand of my hair and left it inside one of the passages, to make a promise that I would return."

I told Sina why I'd come to meet her, about the microbiologists working deep inside the Black Hills, about the subterranean-dwelling bacteria they were finding—ubiquitous creatures, whose potency and significance we were only beginning to understand. I told her about Captain Symmes and our quest to find our shadow-selves, living in the subterranean dark. And I told her about the theory that deep-dwelling creatures were perhaps the first forms of life, that all of life may have originated underground.

"Hmm," she said, nodding, betraying no particular surprise.

Sina took a long pause then began telling me the Lakota creation story.

"The first humans," she said, "lived underground, in the spirit world. The Creator had told them to wait there until the surface world was ready for them. These humans had eyes adapted to living under the surface, where they glowed red and could see in the dark."

Up on the surface, she said, the spider Iktomi was growing lonely. So he packed a bag with all of the most enticing things from the surface—clothing and berries and delicious meat—then he opened a hole in the earth and sent a wolf into the spirit world to deliver the gifts. The humans tried on the buckskin clothing, tasted the berries, and ate the meat, which they especially loved. If they came to the surface, the wolf told them, they would find more meat. The leader of the humans—Tokahe, "the First One"—refused to go and warned everyone of the Creator's instructions to remain underground until the surface was ready for them. But most of the people waved him off and followed the wolf up to the surface. When they arrived aboveground, it was summer, food was plentiful, and they flourished; but as it grew cold, they went hungry. When they asked the Creator for help, he was furious that they had disobeyed him. As their punishment, he turned them into the first herd of buffalo.

"And only then," said Sina, "was the earth ready for humans to live upon. The Creator instructed Tokahe to lead the people to the surface. They made a slow journey upward, stopping to pray four times, the last time at the entrance. When they emerged, the humans followed the bison and learned to survive in the world."

As she finished the story, Sina and I sat quietly at the cave mouth, the hole that Iktomi had opened in the ground. A gust of cool air blew up through the opening, air that had originated deep underground, in the stony womb of the earth.

THE
OCHRE MINERS

What necessity caused man, whose head points
to the stars, to stoop below, burying him
in mines and plunging him in the
very bowels of innermost earth?

—SENECA,

Naturales Quaestiones

In the city of Potosí, people dream each night of the lord of the underworld. It is a city of miners, set high in the icy peaks of the Bolivian Andes, spreading out from the base of a mountain called Cerro Rico, "Rich Hill," which contains some of the richest silver ore in the world. When the first seams were discovered there, in the sixteenth century, thousands of men—all from indigenous tribes who'd been living in the Andean highlands for millennia—came to work in the silver mines. They spent days and nights descending crooked ladders into narrow mineshafts. In the belly of the mountain—at the bottom of sweltering, fetid galleries—they chopped and scraped at the stone walls, and hauled up silver in wooden carts.

Almost as soon as the miners began work in Potosí, whispers started to travel throughout the shafts about a spiritual being who dwelled in the mine, known as El Tío—"The Uncle." He was as magnificently powerful as he was volatile, a spirit who turned from magnanimous to murderous in a flash. It was El Tío who created the silver, and guided the miners to the richest ore. But when his temper turned, El Tío would cast merciless punishment upon the miners, causing toxic fumes to seep from the walls, triggering deadly breakdowns in the dark, knocking miners from their ladders, or striking them with black lung disease. In a city of 150,000 people, every single resident of Potosí had lost a family member to El Tío. Cerro Rico became known as "the mountain that eats men."

On the surface, the miners of Potosí were pious, church-going Catholics; belowground, in the sulfurous dark of the mine, they became worshippers in the ornately sinister cult of El Tío. They climbed down inside the mountain and constructed man-sized statues of the deity in the depths of each shaft. El Tío materialized as a humanlike creature on a throne, with curved horns on his head; flared nostrils; a shaggy, pointed beard; and a large, erect phallus, referencing his profligate appetites. The deity was born of the mine itself, his body sculpted of subterranean clay, his eyes made with discarded mining helmet lightbulbs, his teeth fashioned from shards of crystal.

As the universe of Potosí teetered on the edge of El Tío's tempestuous moods, the miners worked urgently to keep him content. At times, they interacted with him from a delicate remove, tiptoeing around him, as though he might lash out in the dark. It was taboo, for example, to mention God in El Tío's presence, or to make the slightest suggestion that he was anything but omnipotent, lest he grow jealous and lose his temper. Even the sight of a pick-axe, which resembled the Christian cross, might cast him into a paroxysm of fury: when a miner passed before the deity, he would carefully conceal his axe from El Tío's view. Periodically, the miners would load a live llama into a wooden cart and guide it down through the twisting galler-ies, until they reached El Tío, where they sacrificed the animal, spread its blood over the throne, then fed the heart to the deity. As the miners returned to the surface, they prayed that El Tío had eaten his fill, and would not crave human flesh.

And yet, at the end of a long shift, the miners would gather at the foot of the deity, sidling up to his throne like children before an esteemed family elder. In the dark, they would trade jokes and gossip and laugh. Around the circle, they would pass a bottle of *singani* (muscatel grape liquor), pausing to pour a little into El Tío's mouth. They'd place a can of beer in his outstretched clay hand, or feed him coca leaves. When a miner passed a pack of cigarettes around the circle, he would gently place one between El Tío's lips, then lean over to give him a light.

When I first read about the cult of the lord of the mine, I was bewildered. What, I wondered, was the original source of this anxious dance, where the miners dreaded El Tío as the devourer of their families, and yet cozied alongside him in the dark. I scanned anthropology books on the indigenous cultures of the Andes, sifting through ancient religious traditions, but the roots of El Tío seemed to run deeper than any traceable line, as though he were an elemental presence, dwelling down in the earth since long before humans arrived in the region.

Years later, I was reminded of the cult of El Tío when I came across a photograph of a mine dug into a span of hills known as the Weld Range, in the remote outback of Western Australia. Wilgie Mia, as it was called, was the oldest mine in the world, with evidence of Aboriginal visitors going back thirty thousand years. The mine contained ochre, with narrow tunnels that plunged into a rich deposit of the soft, iron-rich, burgundy-colored clay. The photograph showed three men from a tribe called the Wajarri, who had just climbed inside the mine to collect ochre and were now returning to the surface. The photographer had captured them in the midst of performing a peculiar ceremony: upon their emergence, they'd suddenly pivoted, proceeded walking backwards, speaking in anxious whispers, as they used a leafy tree branch to sweep away their footprints. According to the accompanying text, the miners were obscuring their tracks from capricious spirits known as the Mondongs—or, as the photographer called them, "the Devil"—who dwelled in the darkness of the mine and sang mournful songs. The date on the photograph was 1910: a time when white people had only just arrived in the remote parts of Western Australia, when the old Aboriginal traditions were still intact.

Almost from the moment Aboriginal people set foot on

the shores of the continent, more than sixty thousand years ago, they began digging ochre out of the ground. The Aboriginal people regarded the mineral as deeply sacred: for millennia, they traveled on long, highly ritualized pilgrimages in order to climb down into blood-red pits, and dig up ochre from the earth. I wanted to know more about these traditions. I suspected the ritual in the photograph—the anxious wiping away of footprints to avoid disturbing the beast below—might open a keyhole onto an ancient way of thinking about the underground world.

Anthropologists in Australia were quick to tell me that Aboriginal cultures had transformed drastically in the century since the photograph was taken. By the middle of the twentieth century, most of the continent's 250 tribes had been viciously decimated by alcoholism, poverty, and disease, and by violent encounters with white men. Along with so many other Aboriginal customs that had been practiced for millennia, the ochre mining rituals had faded. The mines themselves, once sacred hubs in the Aboriginal landscape, had gone unvisited for years. Some had collapsed into disrepair, some had been forgotten, still others had been devoured by modern mining operations.

All, that is, except for one. I learned of a Wajarri family called the Hamletts, who kept a camp on their ancestral land, at the foot of Wilgie Mia. The family patriarch, Colin Hamlett, had grown up in the Weld Range and had been raised by people who knew Western Australia before the arrival of Europeans, and lived contemporaneously with the miners captured in the photograph. Colin was a head elder of the Wajarri, and a "Traditional Owner" of the Weld Range. In Aboriginal terms, he could still "speak for country," meaning he was one in a dwindling group of elder Aboriginal men and women who still held the ancestral connection to the landscape, who were initiated into the old traditions.

When I reached out to Colin—he wouldn't speak to me directly, rather communicating through a circle of trusted anthropologists and supporters—I learned that his connection to these traditions had lately grown even more urgent, as a modern mining conglomerate called Sinosteel Midwest had recently secured a tenement to build a mine in part of the Weld Range. Just as the velvety ochre had lured Aboriginal miners to the hills for thousands of years, so did the rich iron ore now draw modern industrial miners. As representative of the Wajarri, Colin had spent years fighting tooth-and-nail to prevent the tenements, but had ultimately relented, recognizing that the financial agreement would help the struggling tribe in coming generations. In negotiations, he'd decreed that no Sinosteel drill would come anywhere near Wilgie Mia. It had been several years since the claims were granted, and digging had not commenced, but soon the drills would rumble into the Weld Range. It was for the sake of posterity, I suspected, that Colin invited me to his family's ancestral camp, and to visit the sacred mine, a place white people were seldom allowed. The traditions were still alive at Wilgie Mia: from time to time, just as their ancestors had done for tens of thousands of years, the Hamletts still descended into the mine to collect ochre.

FROM PERTH, ON THE WESTERN COAST, I drove eleven hours
into the heart of the continent, an immense and barren
land known to Australians as the Never-Never. On the
Great Northern Highway, which turned out to be a pinched
two-lane road, I white-knuckled past eighteen-wheeler
after eighteen-wheeler. The land unscrolled on either side,
vast and open on a martian scale. Every few hours, to break
the monotony, I pulled over and wandered into a sea of
gray-green scrub, where I found traces of campfires and
footprints of indeterminate age. In the outpost town of
Cue, I spent a night at the Queen of the Murchison, a
drafty old hotel, where, in the backyard, the owner kept a
fleet of vintage motorcycles, an antique two-decker bus,
and a stand of birdcages containing macaws, their blue and

yellow tropical plumage like alien pennants against the red hills.

THE NEXT MORNING, just after dawn, I headed into the Weld Range. I found the Hamlett camp hidden away in a stand of acacia bushes: a cluster of six or seven Winnebagos, coated in the burgundy dust that permeated every crevice of every surface in the hills. Sitting at a table under a small tarp at the center of camp was Colin, his long silver beard spilling over a protruding belly. He showed his age—he wheezed through absent teeth, had a glaucomal sheen in his eyes—but carried himself with gravity, sat tall in his chair, spine ramrod-straight, with long, sinewy arms crossed over his chest. He emitted charisma and vitality in the form of a physical glow. (The other Wajarri, Colin would tell me, whispered that he was a "featherfoot"—a sorcerer or a shaman—though he assured me it was just a silly superstition.) His family referred to him as "Old Boy," as in "Old Boy keeps an axe and a rifle in the front seat of his truck." In his lap sat a fluffy white dog named Baby, whom he took everywhere.

As he greeted me, Colin gripped my hand just a half second longer than necessary, looking me straight in the eye. It was a reminder, it seemed, that my presence here on his family's land was not ordinary. When I'd told an Aboriginal man I'd met in Perth where I was headed, he'd widened his eyes and told me that the mine was a powerful place. "Watch yourself, mate," he said. "That's big business." The anthropologists I'd spoken with, meanwhile, had advised that I be cautious and deferential. The moment passed; Colin softened and broke into a big operatic laugh. "Look at the city fella," he roared. "Out in the bloody bush!"

The family, seated around the table, had just driven in from the coast, where they lived in the small city of Geraldton, or in the nearby town of Mullewa. It had been some time since they were all together "out country," and despite the specter of the mining tenements and the impending carving up of their land, the mood was light, as beer and cigarettes were passed around. At Colin's side was his wife, Dawn, a round-faced woman with shrewd eyes. Next to them were two of his sons: Carl, known to the family as "Muddy," big, swaggering, with sea-lionish whiskers; and Brendan, or "Cave Man," who was skinny and quiet, with curly hair and large dark eyes. Beyond them was a boisterous gang of nephews and grandsons, most in their twenties. ("Bunch of jackasses," Colin said.) I did my best to follow along, but much of the conversation, which was rapid and peppered with Wajarri, eluded me. I watched two of Colin's grandsons—Kenny and Gordon—fling a lit cigarette back and forth across the circle, catch it between two pinched fingers, take a drag, fling it back.

It was not long before my thoughts turned to Wilgie Mia and the Mondongs. Colin knew why I'd come, of course, but I felt apprehensive about bringing it up, as though an off-key question might cause him to rescind my invitation. I shifted in my chair and craned my neck, trying to figure out where we were in relation to the mine.

"Just on the other side of the trees," said Colin, who'd

evidently been watching me. He pointed his cigarette in the direction of the bushes behind me.

"You'll see her soon enough," he said. Then his face broke into a smile, which somehow managed to convey precisely the opposite of his words: there were, in fact, many things I would have to learn before I was allowed to visit the mine.

"This is an old way," said Colin, indicating the land around us. "It was a long time that blackfellas were coming through to old Wilgie."

FOR AS LONG AS our species has existed, we have dug minerals out of the earth. When *Homo sapiens* first emerged in Africa somewhere between 200,000 and 300,000 years ago, we were already rooting minerals from underground to make tools: flint for blades, basalt for stone axes and hammers, granite for grinding stones. As our ancestors spread across the planet, we dug up every imaginable mineral, from malachite and quartz, to jade and rubies. These stones and metals were always regarded as sacred: worn on protective amulets, endowed with oracular powers, employed in religious rituals. Even when they served utilitarian purposes, they were regarded as agents of transcendence, a means of connecting to the realm of the gods.

Of all the minerals we have rooted out of the earth, none has been revered longer and more universally than red ochre. It was treated as sacred everywhere, from the Andes, where the Inca brushed ochre over tombs, to central India, where hunter-gatherers painted ochre murals on the backs of rock shelters. It has been suggested that the mineral was the first symbol in all of human culture, the first material our ancestors used to gesture beyond the physical world to the metaphysical. It was the discovery of ochre brushed over 100,000-year-old burials in Iraq and Israel that first suggested that early *Homo sapiens* believed in an afterlife. Haifa University archaeologist Ernst Wreschner calls ochre "a red thread" that ties all of humanity together.

Aboriginal Australians pursued ochre with a fervor that baffled early European settlers on the continent. When a missionary compiled a dictionary of a South Australian tribe's language, one of the first phrases he learned was "I crave the red ochre." Homesteaders reported seeing Aboriginal men trek hundreds of miles through the outback, walking for months at a time through desolate landscape, traversing the territory of hostile tribes, just to visit an ochre mine. Upon arrival, they would drop to their knees, kiss the ground, and sob, as though reaching this place had been a matter of life or death.

Red ochre—crushed to a powder, mixed with water, orchid juice, urine, or blood to form paint—was the center of all Aboriginal religious ritual. It was painted on walls of rock shelters in sacred images, some of which remain vivid thirty-five thousand years later; applied to shields before battle; brushed over spears and boomerangs before the hunt. When young men and women were initiated into adulthood, they were painted with ochre; in death, ochre would be painted on the body of the corpse. When archaeologists unearthed the oldest burial in Australia, a man interred on the shores of Lake Mungo sixty thousand years ago, his skeleton was coated with the red mineral.

Ochre connects Aboriginal people to their mythological age of creation, a distant and hazy epoch known as the Dreamtime, when the continent we now call Australia was a formless, pristine expanse. In the Dreamtime, the land was populated by the Ancestors—enormous, powerful animal beings—who moved over the landscape in distinct trails, bringing into existence every hill and river, boulder and tree. Red ochre, it was said, was the Ancestors' blood: wherever there was an ochre deposit in the land, one of the Ancestors had died. To dig ochre out of the earth, to rub it on an object, to paint it on a rock wall, or to apply it to your skin, was to capture the essence of the Ancestors.

That first night at the Hamlett camp, the family told me the Dreamtime story of the creation of Wilgie Mia. One of the Ancestors, a red kangaroo—a *marlu* in Wajarri—was hopping inland from the coast when he was speared by a hunter. As he continued hopping, the wounded marlu began to drip blood, leaving small red splotches on the earth. Gradually, the marlu's hops shortened and grew heavier: each place he landed, a hill sprouted, as blood poured from his wound.

"The old marlu came hopping through," said Colin, popping his finger up and down in the air. "Then he made one last hop." On his final leap, he created the hills, as his innards spilled onto the earth and his blood became the deep red ochre of Wilgie Mia.

In the past, I learned, before you were allowed to visit Wilgie Mia and collect the ochre, you had to first retrace the ritual path of the marlu. In order to teach me the ways of ochre, the Hamletts would not bring me to the mine straightaway; I would have to follow the steps of the old ritual. The mysterious word they used was *songline:* I had to walk the songline of the marlu.

EARLY THE NEXT MORNING, on the very edges of the Weld Range, I headed out across a wide basin rimmed by red cliffs known as Vivienne's Granites. I walked alongside Colin's grandsons Kenny and Gordon and his son Muddy, as well as Muddy's dog, Oscar. The sun was climbing into the sky and we were all beginning to sweat.

"Best to walk with a stick, mate," said Muddy. He spoke in rapid, gravelly bursts, always with one eye shut, giving him the air of a cheerful pirate.

"Why's that?" I asked.

"Dingoes," he said. "And snakes. And *bungarra*," he added, using the Wajarri term for a very large lizard known as a "sand goanna."

The basin, like much of the Western Australian outback, was a desolate, eerily forbidding place. We passed piles of sun-bleached kangaroo bones and red termite hills rising from the ground like spiky red castles; with each gust of wind, red dust devils twisted across the basin floor. So inhospitable was this place, I was surprised when Muddy began to point out traces of ancient visitors. At first, a few small petroglyphs carved into the face of a boulder; then a handful of stone flakes on the ground. Then, as my eyes adjusted to the landscape, it became clear that we were wading through a whole field of grinding stones, broken stone axes, ancient campfires with ashes full of emu shells, even a waterhole with edges worn by millennia worth of visitors kneeling to drink. In the past, this place had been a hub. The marlu ancestor, as the story went, hopped across this basin on the route to Wilgie Mia.

"The *yamaji* came through here," said Muddy, using the Wajarri word for "people." "We're on the songline now."

The songlines—a term originating with anthropologists in the 1940s, but popularized in the 1980s by the English travel writer Bruce Chatwin—are an enigmatic spiritual system passed down through Aboriginal culture over untold thousands of years. A songline is a path marking the trail of a Dreamtime ancestor—an emu, a wallaby, a dingo, a marlu—as they moved across the primordial continent, bringing the landscape into existence. Thousands of songlines crisscross Australia like strings in a giant net, running through empty desert outback, along rocky seacoasts, down into shaded forests, with some stretching clear from one side of Australia to the other. A songline is a physical path, like a route on a map, which helped the Aboriginal people navigate between sacred landmarks; at the same time, in a way that confounds modern Western conceptions of time and space, a songline is a story, recounting the saga of the ancestor's sacred journey. The best analogy would be if the Bible or the *Iliad* or the Mahabharata were not books, but collections of paths in the earth, and instead of reading them on a page, you walked their length, singing out the story, the rhythm of the tale matching the meter of your gait and the contours of the land. Songlines are seldom discussed with outsiders, and while the Hamletts had agreed to show me the songline of the marlu, much would be left unsaid: some knowledge was lost, some too sacred to share.

The ritual pilgrimage on the songline to Wilgie Mia would have unfolded over several weeks, like an extended choreographed dance. The ochre party—a small group of men, as women were prohibited from entering ochre mines—would embark from a great distance, perhaps hundreds of miles away. Falling into step with the songline of the marlu, they would sing out the story of the ancestor, giving a real-time account of every trial and adventure. Nearer to the mine, as the party began to see traces of ochre in the landscape, streaks of red pigment left by past ochre expeditions, the ritual would heighten, the ceremony growing tighter and more elaborate.

A 1904 account of an ochre expedition to a mine called Yerkinna in southern Australia provides a vivid description of how the approach to Wilgie Mia would have transpired. The members of the party—men from a tribe known as the Kuyani—walked the songline over the course of five weeks.

Before their final approach, they fasted, giving up food and water, then shaved their bodies of every last hair and greased their torsos with the fat of an iguana. On the final night, they spent the whole night dancing, with no sleep allowed, until dawn, when they broke into a sprint to the mouth of the mine. The consequences of failing to perform any of these ceremonies were dire. In the 1870s, an ochre party visiting the Yerkinna mine failed to follow the precise protocol: as they entered the mine, the roof collapsed and the entire party, save one man, perished, entombed in red ochre. In the aftermath, it was said that the guardian spirits of the mine—versions of the Mondongs—had been insulted and had taken their revenge.

WE HEADED TO THE outer ridge of the basin, trailed by a haze of flies as we walked along the cliff line. Oscar darted after lizards in the bushes. A white owl swooped out of a crack in the cliffs with a startling drum of the wings. We had not been walking long when Gordon dropped to all fours and crawled down into a small alcove, gesturing for me to follow him.

"Holy shit," I said, crouching in the gloom. I could hear Kenny and Muddy, outside the crevice, laughing out loud.

At some point in the past, an Aboriginal person had placed their hand against the rock wall and blown ochre over it through their mouth, creating a stencil. I inched up close.

"That's the marlu blood," said Muddy.

Up ahead, we crawled into another alcove, found another ochre handprint. In the next niche, two more prints, one with a crooked thumb. Soon we were running down the cliff line. We found dozens of prints. Under a bush, there was a grindstone stained with ochre. On a wall beneath an over-hang, an ochre stencil of two boomerangs, with rounded edges facing each other. Everywhere were signs of visitors passing up and down this route, their every movement traced in ochre, like a soft red brushstroke over the land.

At one point, shoved into a nook in the rock next to an ochre print, I spotted a bundle of twigs, elaborately woven into the shape of a bird's nest. As I crawled forward to get a closer look, Kenny quickly whispered, "Better not touch that."

When I turned around, Kenny, Gordon, and Muddy were looking at me with silent gravity.

"That's blackfella business," said Muddy.

Later that night, as we sat around a campfire, sur-rounded by a low thicket of acacia bushes, Kenny told me about the bundle of twigs. A few years before, he said, his cousin Brian had been exploring the cliffs near the ancient mine, when he came across a similar bundle of twigs, entwined in a tight weave. To get a closer look, he pulled it out of the alcove, turned it over in his hands; he then tried to replace it, but it was clear he had disturbed the relic. That night, Kenny told me, Brian fell ill and he was taken to the hospital. He stayed in bed for three weeks.

"Was it the Mondongs?" I asked. But even as the question came out of my mouth, I swallowed the last word, immediately feeling crude and foolish for saying it out loud.

Kenny did not respond or give any indication that he had heard me.

FOR AS LONG AS we have extracted minerals from the earth, mining has been a spiritual act, accompanied by elaborate ritual and ceremony. Throughout the ancient world, cul-

tures spoke of stones and ores hidden underground as em-
bryonic outgrowths of the larger earthly body. Across the
slow progression of geological time, they gestated and grew,
ripened and matured down in the warm earth. In ancient
Mesopotamia, for example, the Assyrian word for "mineral"
was *ku-bu*, which also translates as "fetus" or "embryo." The
Cherokee, meanwhile, nurtured their crystals as sentient
creatures, and fed them with the blood of animals. In turn,
the act of rooting minerals out of the earth was considered
a spiritual transgression, akin to ripping the entrails from
inside of a body. The moment you climbed underground
with a digging tool, you were intruding into sacred mysteries, engaging in an act of acute spiritual anxiety.

Virtually every mine in the pre-modern world was
haunted by some form of earth spirit: a capricious being,
sometimes benevolent, but more often vengeful. In Ukrainian mines, it was Shubin, a spirit in a long fur coat, who
would lead miners to the richest veins, or trigger a deadly
breakdown. Germanic miners whispered of vindictive
gnomes and trolls who made minerals in the dark glitter so
brightly, they'd blind any man who approached. In Britain,
it was two-foot-tall men called Knockers, who would tap
on the walls, drawing the miners underground, only to release hazardous fumes from the walls. No one dared dig
stone or ore from the earth without first engaging in a
painstaking negotiation with these beings. Just as the silver
miners in Bolivia plied El Tío with llama hearts, and Aboriginal people walked the songlines, miners all over the
world performed elaborate ceremonies of appeasement.
Priests and shamans oversaw the groundbreaking of mines;

shrines and temples were constructed at mine entrances; animals were slaughtered as offerings. In the Mande culture in West Africa, a miner would sequester himself from the rest of society for several days, fasting and observing sexual abstinence in order to purify himself, before climbing down to dig into the earth.

If ancient miners saw our industrial-scale mining practices of today, where machines gouge mega-pits out of the earth, they would surely recoil. They would accuse us of entering recklessly into a delicate transaction with the earth, of courting tragedy and catastrophe. Each time a mine collapsed, entombing hundreds of men, each time a fire ripped through the subterranean galleries and burned miners alive, each time the chemicals of a mine poisoned a river or sent disease through a local community, they would point to our sacrilege, our failure to appease the spirits.

IT WAS LATE, the stars were out, we'd just finished eating a kangaroo stew. Dawn had insisted on giving me the tail, the most tender part. The remainder of the animal, which Colin had shot from the cab of his pickup a few hours before, was slung from the branches of a nearby tree. Everyone leaned back in their chairs and passed around cigarettes. Muddy rubbed his belly and whisper-sang, "Sweet mar-lu," drawing out the last syllable: *looo*.

Colin had informed me that I would visit Wilgie Mia the following morning. He was getting too creaky, he said, for the steep climb into the mine, and so my guide would be Brendan. Of his sons, Colin said, Brendan could best speak to country. In his free time, he carved spears and boomerangs; when the Wajarri gathered as a tribe, it was Brendan who led the dances.

Now seated next to me, Brendan leaned over and passed a pouch of rolling tobacco. "If you're real quiet in the mine tomorrow," he said, speaking so low that I had to edge forward to hear him, "you'll hear the old Mondongs singing." He quietly imitated the song: a low, plaintive wail, vibrating from the back of the throat.

Everyone was quiet again, until Colin spoke. "The Mondongs," he said, a half smile just visible under the brim of his hat, "look like old blackfellas, only they're smaller. They come out stark naked. They appear and disappear quick."

From there, the Hamletts began going around the circle, volunteering stories about the Mondongs. Dawn told of an anthropologist who was out working near Wilgie Mia when an old man—small, dark-skinned, naked— appeared up on the lip of the mine and glared down at him, began singing a haunting song. The anthropologist got into his car, sped away, and never returned.

Colin told a similar story about a female anthropologist who claimed that an old man appeared before her, told her to leave the place, then demanded that she wash away her footprints from Wilgie Mia. She, too, left the hills and never returned. Colin and Dawn now began laughing softly, as the sons and grandsons quietly started to chuckle.

Muddy told of the time his wife unwittingly brought a Mondong back to town in her car. People asked her about the little old man sitting in the passenger seat—only, she hadn't picked anyone up. At this, the whole family lost it, laughing uproariously, spitting out beer and leaning back in their chairs, slapping the table.

I listened, confused by the tone of the stories, unsure of what to say. For all the Mondongs' supposed menace, the Hamletts spoke of them with warmth and nostalgia, as though recounting old family lore.

Colin managed to suppress his mirth just long enough to tell a story from a few years prior, when a group of Wajarri people came to visit Wilgie Mia as part of a community meeting. They all went underground to visit the sacred mine during the day, but as soon as the sun went down, and darkness fell, no one would go near the site. The group made camp about a mile away from the mine, all of them trembling in the dark, terrified of the Mondongs. All of the Hamletts were now shrieking, practically falling out of their chairs, drying tears from their eyes. "All of these scared blackfellas," Colin howled, "going to piss in pairs!"

It slowly dawned on me that I was hearing the same anxious back-and-forth as with El Tío at Potosí, where dread mingled with an almost familial intimacy.

The laughter soon subsided, and Colin went quiet, his face darkening under his hat. "They're bloody there, all right," he said, his voice flat, even taking an edge. He took a drag from his cigarette and looked at me. "You just have to know how to handle them."

THE FOLLOWING MORNING, under a soft dawn mist, we bounced through the bush in Colin's pickup and arrived at

the foot of Wilgie Mia, which rose out of the earth abruptly and cinematically, like a fever-red volcano. Brendan and I gathered our gear and lights from the truck while Colin and Dawn, along with Baby, set up chairs and a thermos of coffee in the shadow of the car. On the ground all around us were flakes chipped from stone tools, remnants of thousands of years' worth of visitors to the mine. No one said much. Colin, with a note of warmth that surprised me, gave me a wink. He had insisted we come just after dawn, when the colors of the hills would be at their most vigorous.

As Brendan and I started up the slope, and Colin's truck receded below, a view of the entire Weld Range opened up all around us. The route of the songline wended through the hills, each peak marking where the marlu ancestor had hopped. We clambered over red-and-black marbled outcroppings that contained the purest iron ore in the whole

Weld Range—the material that Sinosteel would love to mine. "They'll never touch it," said Brendan. "Better chance of pissing on the moon."

He reached the top of the hill first. "Here she is," he said quietly. I climbed up next to him, and stared down into a gaping red hollow, which dropped deeper into the earth than I could see. The color left me stupefied. The entire landscape of the Weld Range expressed vibrant colors—magenta canyons at sunset, crimson puddles in the rain—but this was a whole other animal of color. It was the red of lava, the red of the womb—it felt like the place where red began.

And perhaps it was the visceral shock of the color, or the curious animal warmth emanating from deep within, or maybe I was just bleary in the early dawn hour, but for just a moment, as I peered down over the lip of the mine, I could've sworn that I saw something stir in the gloom, a small, spritelike man, winking in and out of view.

Brendan climbed over the lip and I followed. Down the steep pitch, I resorted to a lurching, feet-first crab walk. Powdered ochre fell away at our feet in long, hissing cascades. Within seconds, it seemed, my whole body was stained red, like a baptism. As we climbed down, the softness of the ochre absorbed all sound, giving our movements a dreamlike noiselessness: when we spoke, our voices sounded cottony and remote, as though they were coming from the mouths of other people. As the sun rose, and light angled down through the mouth of the mine, the ochre shimmered and changed colors from warm burgundy to electric violet to a searing soprano pink. It gave the walls the illusion of movement, as though the entire mine were gently pulsating. We were traveling down the throat of a living creature, being swallowed by the earth.

We paused for a moment and Brendan picked a clump of ochre out of the walls. Even as he handed it to me, I felt nervous about touching it.

"It's okay," he said. "Old Boy wanted me to find you a good piece."

The ochre was softer and lighter than I'd expected. I closed my hand around the clump and it instantly dissolved into a perfect powder, the consistency of blush on a woman's cheek. I pressed it between my hands, rubbed it over my fingers—it glowed in my palm.

Halfway down, we paused on a ledge where we were directly on the edge of the twilight zone, with the light from the entrance still visible above, and pitch-darkness at our feet. A dense aroma of bat guano rose from out of view.

Lying next to us on the ledge, I noticed, was the carcass of a kangaroo, its brittle skin stained deep purple—the very marlu we'd been telling stories about for days.

"They hop down here looking for water," Brendan said, "then they can't get out."

Before we went farther, Brendan left me for a moment, and disappeared into a dark corner of the mine, then reappeared holding a wooden stick. It was bone-colored, ancient, smooth with years and years of handling. While exploring the mine one day, he said, he'd found it tucked away in a crevice, where it seemed to have been hidden intentionally. He believed it was a digging stick, a mining tool, used by his ancestors to dig ochre from the walls.

He restored the stick to its hiding place, and we continued our descent, down into the mine's dark zone, where the Mondongs sing.

Brendan motioned for me to click on my headlamp. "Hope you're not claustrophobic, mate," he said.

As we proceeded into the dark, we fell into step with the old miners. The ritual would have been performed by a small clan of specialists—ochre priests—who were uniquely initiated into the laws of ochre. The visiting party, having walked the songline, having told the dreaming story of the marlu, having "opened the way," would be escorted by the priests down a long tunnel and into the red heart of the mine. In the past, the mine's roof, where Brendan and I had entered, had been closed, except for a small aperture, which would have let in only a narrow shaft of light. While the visitors waited, the ochre priests would descend into the deep parts of the mine, following the mineral seam, each gripping a digging tool, perhaps the very stick that Brendan had found hidden in the walls. Deep underground, they would have softly and delicately chopped at the walls. They would have gathered the fallen ochre, mixed it with water, and rolled it into large balls to be delivered to the visitors. As the ochre priests made their way back to the surface, they would have turned and walked out of the mine backwards, using a leafy branch to sweep away their footprints, covering their tracks from the Mondongs.

We crouched and ducked and crawled through tight quarry tunnels, both of us saturated in ochre dust, down to our fingernails and eyelids. It grew hotter and hotter, the air growing short, the stench of guano more robust. I heard the whisper of wings as bats swooped back and forth above our heads. In the walls, I saw scratch marks where miners in the past had dug out the ochre.

We paused for a moment, sitting with our backs against the tunnel wall. I felt Brendan next to me in the dark, both of us tense and silent, both of us listening for the Mondongs.

After several moments, Brendan shook his head. "Not today," he said. "They're leaving us alone."

I nodded. I would not hear the song of the Mondongs.

But down in the dark, enveloped in the hush of ochre, I could feel their presence—just as I could feel the presence of El Tío, and every other ancient earth spirit who'd once guarded mines all over the world. That is to say, I could feel the specific anxiety from which these spirits were born. The dissonance of climbing down inside a hallowed place with tool in hand, where we are chopping at the earth, but also chopping at a living body, chopping in order to extract ancient material, something mysterious and sacred and rare, a substance beyond our world, and trying to take that strange matter from dark into light.

EARLY THE FOLLOWING MORNING, after having coffee with Colin and Dawn and Baby in their trailer, I made my way out of the Weld Range and headed back to Cue. On the way, I passed a large rumbling truck headed into the hills. It was a Sinosteel truck, one in a larger fleet, now en route to a small camp that the company had already built in the range, on the other side of the hills from Wilgie Mia. The company was eager to break ground: a long time had passed since the claims were granted, and now the operation was many years behind schedule. Sinosteel had encountered delays and setbacks at every turn: funding had been pulled, infrastructure had collapsed, local politicians had scrambled their mission. No one doubted that drills would one day rumble up this road and begin carving up the earth, but as I drove out of the hills, with red ochre still stained in the creases of my palms, I imagined the Mondongs flickering through the Weld Range, doing their best to thwart the miners who had not followed the laws, who disrespected their stewardship of the hills and had lost touch with the old ways of the earth.

CHAPTER 5

THE BURROWERS

When it comes to excavated ground,
dreams have no limit.

—GASTON BACHELARD,
The Poetics of Space

One day in the early 1960s in northeast London, a man named William Lyttle set out to dig a wine cellar in his basement. Lyttle—a wiry, sharp-chinned man who worked as a civil engineer—brought a shovel down beneath his home and began digging into the walls. For several hours, he scooped out damp earth and cast it behind him, until, eventually, he'd managed to excavate a hollow of sufficient size for a wine cellar. But then Lyttle didn't stop. Perhaps he enjoyed the rhythm of the work, the chop of the shovel, the smell of clay—or perhaps it was something else entirely. In any case, Lyttle kept digging. And digging. For *forty years* he dug.

Lyttle's neighbors in the neighborhood of Hackney watched as he trucked wheelbarrows of debris up from his basement, dumping it into a mound in his backyard. At first, they joked that he was building an underground swimming pool, but as the years passed, and Lyttle kept digging, the jokes faded. As the mound in his garden grew, the house fell into neglect: broken windows went unfixed, vines crept up the façade, and sections of the roof collapsed. Lyttle always wore the same filthy suit jacket and grew out a wolverine-ish beard. Neighbors claimed they could hear him beneath their gardens at night, scratching away like an animal.

In 2006, the sidewalk in front of Lyttle's house caved in. Representatives from the city council came to investigate and found themselves wandering through an enormous warren of earthy tunnels in the basement. It encompassed multiple levels, reaching thirty feet down, radiating out sixty feet in every direction. Some tunnels were low and narrow, others large, braced by household appliances stacked one on top of the other. One visitor at the time commented that Lyttle had dug out his basement into "a giant ant nest." The house was deemed uninhabitable and Lyttle was relocated to an apartment in a high-rise owned by the council. They installed him on the top floor to preclude the temptation to burrow.

When the press caught word of the tunnels, Lyttle was briefly thrust into the spotlight, as the tabloids ran stories about the "Mole Man of Hackney." Pubs in the neighborhood celebrated him as a local hero, while Londoners made pilgrimages to see Lyttle's house, now braced in scaffolding.

On the façade, the city installed a plaque: WILLIAM "MOLE MAN" LYTTLE. BURROWER. LIVED AND DUG HERE. Upon Lyttle's death, in 2010, members of the council entered the apartment in the high-rise, only to find that he'd started digging holes into the walls, burrowing from one room to the next.

I came across Lyttle's story shortly after he passed, just as his strange, perforated house was put up for auction. As I studied the interviews he'd given over the years, I realized that Lyttle had never given an explanation of what com-

pelled him to burrow. "I guess I'm a man who enjoys digging," he said to one reporter. "I just wanted a big basement," he told another. In one interview, he said, "There is great beauty in inventing things that serve no purpose." I found an old photograph of Lyttle emerging from debris in his yard, looking filthy and feral, but with an expression on his face that was almost beatific, as though the burrower were in possession of a precious secret.

William Lyttle, I learned, was not alone. All over the world, there were people who, for reasons they could not quite articulate, had fallen into a kind of fugue state and dedicated their lives to digging underground—a whole case file of Mole Men. There was Lyova Arakelyan, a man in rural Armenia who, while excavating a potato cellar beneath his home, became transfixed, and spent the next three decades digging winding tunnels and spiral staircases. To those who asked why, he only explained that each night he heard voices in his dreams telling him to dig. And the entomologist Harrison G. Dyar, Jr., who excavated a quarter mile's worth of tunnels beneath two separate houses in Washington, D.C. When the tunnels were revealed in 1924, after a car fell through the street, Dyar told the press, "I do it for the exercise." And an old man in the Mojave Desert, William "Burro" Schmidt, who spent thirty-two years pickaxing a 2,087-foot-long tunnel into the side of a solid granite mountain. ("Just a shortcut, I suppose.") And a young man named Elton Macdonald, who covertly excavated a thirty-foot-long tunnel beneath a city park in Toronto, which caused a city-wide panic after the police announced the tunnel as a potential hideout for terrorists.

When Macdonald revealed himself as the burrower, he could only explain, "Digging relaxes me." And then Lord William Cavendish-Scott-Bentinck, a nineteenth-century duke, who, along with a crew of laborers, hollowed out an entire tunnel metropolis beneath his estate, complete with an underground library, a billiards room, and a ten-thousand-square-foot underground ballroom made entirely of clay, which the duke used as a private roller-skating rink.

As I came upon one after another of the Mole Men, I began to imagine a whole new psychological syndrome. A new entry in the DSM-5: *perforomania*, from the Latin *perforo*, meaning "to dig, to tunnel, to burrow." In any case, I suspected the Mole Men were only the tip of a much broader and deeper-rooted impulse.

I FIRST READ ABOUT the burrowers of Cappadocia in an old guidebook for Turkey. It described a thin scatter of villages and small cities across an enormous plateau at the heart of the country. The region was underlain by tuff, a soft stone formed from deep deposits of compacted volcanic ash. As rocks go, tuff was like Play-Doh: easily manipulated, but firm enough to hold its form, which is to say, prime burrowing material. If marble statuary had reached its height in Renaissance-era Florence, burrowing had found its apex in ancient Cappadocia.

Beneath almost every settlement in the region, I read, was a hand-carved network of caverns, connected by sinuous tunnels: the guidebook called them "underground cities." Some were enormous, like upside-down castles, reaching more than ten stories into the earth, with space to hold thousands of people. And there were hundreds of them throughout the region. Archaeologists understood that at least sometimes the underground cities were used for refuge: when enemies attacked, the inhabitants of a village on the surface would retreat underground for safety. Beyond that, though, the cities were inscrutable. The structures were unmentioned in ancient texts and yielded little in the way of archaeological material. Some could be traced back to early Christians, who lived in Cappadocia during the third and fourth centuries, but others went back much further, into foggier reaches of prehistory. One story traced them back to an ancient Persian king called Yima, who, in order to save his kingdom from a coming disaster, excavated an enormous underground shelter, with multiple levels and winding tunnels. What caught my eye in the guidebook was a small illustration, a cross section of one of the cities, revealing an astonishing warren, full of people digging into the earth. Here was an entire culture—

forebears of William Lyttle and the Mole Men—who had dedicated their lives to burrowing.

I ARRIVED IN CAPPADOCIA by overnight bus from Istanbul. As I disembarked in Göreme, a town in the heart of the region, I blinked dumbly at the landscape: millions of years of wind and rain had sculpted the tuff into creamy stone hills that felt geologically implausible, like something from a child's illustration of an alien planet. Just beyond the bus station was a cluster of upright stone obelisks called *peri bacalari,* or "fairy chimneys": depending on whom you asked, the fairy chimneys radiated good luck, or they were inhabited by vindictive elves and should be avoided at all costs.

My base was Emre's Cave House, a small cluster of rooms scooped out of a stone hill on the outskirts of town. It was the cheapest place around and a bit forlorn, with a rusted swimming pool in the front lawn. The eponymous manager was a large-bellied, melancholic man who drank red wine and showed the female guests pictures of his horse, which he kept on a nearby farm.

Each day, I set out from Göreme to visit a different underground city. A few of the sites were linked by bus routes, but most were in remote villages. I'd hike down long, winding roads, where truckers and farmers would pick me up. The names of the villages—Özkonak, Derinkuyu, Kaymakli—were as craggy as the land itself. I carried a textbook on local archaeology that I'd picked up in Istanbul, written by a historian called Ömer Demir, whose ardent love for the underground cities, along with his idiosyncratically translated English, made him an entertaining sidekick.

One foggy morning, I arrived in Özlüce, a small village in the middle of a volcanic sweep. Hiking up from the main road, I encountered an old woman in a kerchief and billowing pants. She was sweeping the doorway of her home as gray smoke curled out of the chimney above her head.

I said, *"Yeralti Sehri?"* which is Turkish for "underground city." (The extent of my Turkish: "hello," "how are you?" "thank you," "goodbye," and "underground city.") She beckoned for me to follow her and led me to a small building, which turned out to be conveniently marked.

On the other side of the door, I clicked on my lamp and descended a stone-cut stairway, down into a hive of pitch-dark caverns. The walls were caramel-colored, the air damp and cold enough that my breath was visible. I proceeded slowly, crawling on hands and knees through low passageways, squeezing through narrow corridors, ducking beneath arches.

There were maybe a half dozen chambers, ranging from closet-sized to four-car-garage, all of them linked by narrow tunnels. Everything was rough-hewn: no hard angles, only soft amoebalike shapes. The chambers were full of dust and cobwebs and smelled of mildew—it had been a long time since anyone had stepped foot in here. After a half hour or so, I found that I had a hard time reading the space: it felt cold, alien, and empty. I saw no sign of the collective digging I'd seen in the cutaway illustration in the guidebook. And no hints, certainly, of the roots of our impulse to burrow. I was looking in the wrong place, it seemed, or perhaps looking at the space in the wrong way.

As I made my way back to the entrance, I saw, to the side of the doorway, a giant, disc-shaped stone: the size and shape of a monster-truck tire, weighing several thousand pounds. It sat upright on its edge, tucked into a slot. In the event of an invasion, according to Ömer, the villagers would retreat underground and roll the stone in front of the door,

sealing the city from inside. The millstones guarded the entrances of all the underground cities in the region.

In each city I visited, I made maps and diagrams, recorded inventories of objects I discovered, photographed every tunnel and chamber, and ran my hand over each millstone. Sometimes I stayed underground for hours, until my fingers went numb. However long I searched, I always felt that I was misreading the space, as though there was something in the lineage of the underground cities that eluded me. Sometimes, after hours of exploring, I'd sit in a deep chamber and knock at the walls with my knuckles, listening for variations in the sound, searching for a hidden chamber that might hold a clue.

One afternoon, in the village of Özkonak, I met an old farmer named Latif who had himself discovered an underground city. Latif was the town imam and spoke in a deep and resonant voice; he had only one arm, having lost the other after he fell from a tree as a child. One day in 1972, he told me, he was walking his fields when he noticed water disappearing underground. He started prodding, a hole opened up, and he felt cool air against his face. He continued digging; one chamber led to another, which led to another, deeper and deeper into the earth. I asked Latif how it felt to make such a discovery, to unearth such a bewildering structure. For a moment, he looked at me thoughtfully, clicking prayer beads between his fingers, then he said something that took me entirely by surprise. "The underground cities are not so strange," he said. "They are everywhere. To dig these spaces is a very old matter. It is a natural thing to do."

———

IN FACT, THE VERY first complex animal in the history of life was a burrower. *Ediacaran fauna* were tiny, enigmatic creatures who lived 542 million years ago: our first oxygen-breathing, multicellular organisms, the first entrants into what paleontologists call the Phanerozoic, or the "Eon of Visible Life." They dwelled on the ocean floor and dug webs of tunnels to protect themselves in the earth. Fossils of their burrows—beautiful, ghostly designs called "traces"—have been found by paleontologists in every corner of the planet.

Since then, burrowing has become one of evolution's most vital modes of existence, a way for creatures to thwart predators, to protect their young, to shield themselves from the elements. Animals in every precinct of life and every habitat have thrived as burrowers, from fish who dig beneath the sea floor to birds who burrow into the desert. Indeed, the creatures known to biologists as the "most successful terrestrial animals in the history of life" are burrowers: that is, the ants. Having thrived in every part of the planet for 100 million years, ants make up roughly 15 percent of the entire biomass of terrestrial life. They have long operated out of enormous, ingeniously designed subterranean nests, some of which run thirty feet deep and cover the area of a small house, encompassing hundreds of entrances and thousands of chambers, each dedicated to a specific function: some for storing food, some for disposing of waste, some for raising the colony's brood.

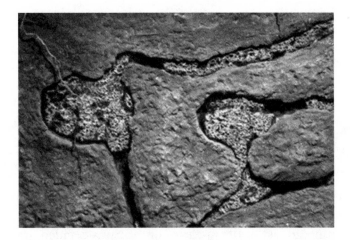

By all evolutionary logic, humans should not burrow. We are oversized, upright, long-limbed; our livelihood depends on abundant air and light. Physiologically speaking, there is no environment so intolerable as a tight, dark, underground enclosure, where oxygen is scarce. To burrow is to experience claustrophobia in its most crystallized form, like enclosing yourself in a tomb.

And yet, throughout history, in every corner of the world, we have burrowed. In times of war or conflict, when we are at our most desperate, we have dug down into the dark, and enclosed ourselves in the "very thickness of the planet," as the philosopher Paul Virilio writes in his study of subterranean refuge, *Bunker Archaeology*. The people of Malta, during the sixteenth century, excavated labyrinthine networks beneath their cities to ward off the invading Turks; just as the Viet Cong carved out spidering tunnel cities beneath the jungle floor; just as billionaire tech lords in Silicon Valley now excavate monumental luxury bunker complexes in anticipation of the apocalypse. It's one of history's oldest stories. The prophet Isaiah describes the day the Lord delivers his wrath upon the heretics: "For fear of the Lord, and the glory of his majesty, [they] shall go into holes of the rocks, and into the caves of the earth."

The greatest frenzy of burrowing of the modern era occurred in the United States, during the Cold War. The Russians and the Americans—like "two scorpions in a bottle," as J. Robert Oppenheimer wrote—were twitching fingers over missile launch buttons. The only way to survive the imminent blast of the nuclear bomb, it was decided, was to burrow underground.

In suburban backyards, families took up shovels and dug fallout shelters and foxholes, stocking them with water drums and survival biscuits, known as "atomic crackers." Hundreds of companies offered ready-made personal shelters, which, like Winnebagos or Jacuzzis, came in multiple models, from "economy" to "luxury."

The town of Artesia, New Mexico, built a subterranean school. The only visible aspect was the playground: beneath the surface were classrooms for 420 students, and, in the event of a nuclear strike, shelter for 2,000 citizens. The cafeteria's walk-in refrigerator could be converted into a morgue. One student told a newspaper reporter that "being underground gives you a funny feeling, but you know you're safe."

In New York City, meanwhile, the government weighed a proposal for the Manhattan Shelter Project, an 800-foot-deep shelter, built to house the entire population of Manhattan—four million people, at the time—for up to ninety days. The proposed shelter had ninety-two access points, ensuring that all Manhattanites could be inside the blast doors within thirty minutes.

"Never in America," wrote one journalist at the time, "has so much dirt been excavated, by so many, so feverishly." *The New York Times* described this scene: "Last week, a six-year-old boy was found busily digging a hole in the smooth lawn in front of his home. 'What are you doing?' his horrified mother asked. Without stopping, the boy answered, 'I'm digging a big hole in the ground to hide from the bomb.'"

Critics at the time argued that digging down into the

earth was unnatural: an animal impulse, rather than a human one. "Our grave-digging acquiescence," as one writer called it, inverted the trajectory of our species. "When primitive man left his cave and traveled into the light," he wrote, "he was meant to travel onward and upward, not to circle back."And yet, there we were, gripping shovels, casting great showers of dirt into the air: all of us caught in the same trance, it seemed, as William Lyttle, burrowing beneath his home.

FOLLOWING MY CONVERSATION WITH LATIF, I visited an underground city called Derinkuyu, which was the largest of all of the underground cities. At the center of a wind-swept field, I descended a long stone-cut staircase going down thirty feet below the surface. From the moment I passed the entrance millstone, I felt a strong breeze blowing up from below, a sign of a vast and deep network.

I wandered through one chamber—a stable for live-

stock, according to Ömer—then another, before I arrived
in a large chamber that had once been a kitchen. There was
a small pit in the middle of the floor that would have held
a fire for cooking and alcoves in the walls to hold candles;
in the next chamber, a pantry for earthenware jugs of grain.
There were openings in the ceiling for ventilation, where
cool air came rushing down, and deep wells plunging to the
water table. Farther on, I passed a dormitory, followed by a
large chamber that, according to Ömer, had been used as a
schoolroom. Only a small section of the underground city
of Derinkuyu had been cleared out and made passable to
visitors. In the past, there had been as many as eighteen
levels, with hundreds of chambers and ventilation shafts,
and over forty entrance points, most of them now obscured
beneath modern buildings.

   As I wandered through Derinkuyu and saw just how
tubular and knobby the tunnels were, and how vast and

sprawling the network, I found that the scale made me feel like I'd been zapped into miniature. Moving from chamber to chamber, I found myself imagining that at any moment I might turn down a passageway and be swept up in a wave of ants, rushing through the dark.

It was a feeling that only continued to resonate as I climbed back up to the surface, where, not far from Derinkuyu, I walked down a dried-up river gorge. Erosion had caused a part of the underground city to collapse, leaving

an exposed cross-section view. It stretched for almost a mile, revealing a perfect X-ray of subterranean architecture.

As I slowly made my way down the gorge, I couldn't help but notice that the cross section of the underground city bore an uncanny resemblance to the cutaway view of an ant nest.

As I contemplated this curious architectural reflection, walking deeper down the gorge, it began to rain. I ducked over to one bank and huddled beneath the lip of a chamber in the ancient underground city, where I watched raindrops fall on the dusty ground in front of me. I recalled the line from Democritus about how "we are pupils of the animals in the most important things." I wondered if the ant and human echo was the result of some form of teaching, a diffusion of ideas between species. I though of an old myth told by the Hopi tribe, which holds that in the deep past, a tremendous fire broke out on the land and nearly wiped out the humans, until they were rescued at the last moment by the ants. As the fire bore down, the ants arrived and led the humans into their nests, where they kept them safe in the tunnels beneath the earth, until the fire calmed. When the humans emerged on the surface and began rebuilding their lives, they remained forever bound in gratitude to the ants.

In any case, the convergence stayed with me, and when I returned home, I took a trip to Tallahassee, Florida, to visit an entomologist who studied the architecture of ant nests.

WALTER TSCHINKEL PICKED ME UP on a steamy morning in Tallahassee and we drove out to his research station in the Apalachicola National Forest. He was in his late sixties and had been studying ants for a half century. Along the way, we talked about his childhood in Alabama, where he grew up exploring the many caves near his home and following ants, but conversation dwindled and soon we were riding in silence. I could see that he was a taciturn, no-nonsense man, and I held off on telling him the entire story of why I'd come, about the Mole Men and the underground cities, and my questions about our impulse to burrow.

Tschinkel's research station was a sandy clearing enclosed by scrub trees, which contained nests of two species of ant: *Paratrechina arenivaga* and *Aphaenogaster floridana*. In order to locate active nests, we staked pieces of Vienna sausage to the ground and waited for the ants to emerge. In his years of studying nest architecture and trying to understand how ants used the various parts of a nest, Tschinkel had been frustrated by the fact that you could never really *see* a nest, because in the act of excavating it, you destroyed it. His solution was to make metal casts of the nests.

In a homemade kiln, we melted down scraps of zinc, which he'd salvaged from old anodes at marine shipyards. Then we put on heavy oven mitts, carried the crucible over to each of the nests, and poured the molten zinc into the entrances. The hot silvery liquid pooled and disappeared underground: the nest's inhabitants were, unfortunately, sacrificed. "Death," said Tschinkel, "is part of biology."

Next to one of the nests, we dug a large pit in the ground. The zinc had flowed down into every artery and chamber and node, then hardened. We gingerly picked the mold out of the earth, watching it emerge from the soil like a strange relic from an ancient civilization.

Later, Tschinkel added the casts we'd made to his collection. He displayed it in his garage, where dozens of metal ant nest molds hung from the ceiling like metallic chandeliers. Each, he explained, was the creation of a different species of ant.

Some of which could get to be quite large.

But as I picked up one cast, which was a nest of *Aphaenogaster floridana,* one of the species whose nest we'd cast in the forest, I had the uncanny feeling that I was holding a precise miniature model of the underground city of Derinkuyu.

We had spent the day working mostly in silence, but now I was no longer able to contain myself and I started telling Tschinkel all about why I'd come to see him: about William Lyttle and about the Cold War burrowing and about the underground cities of Cappadocia and the un-

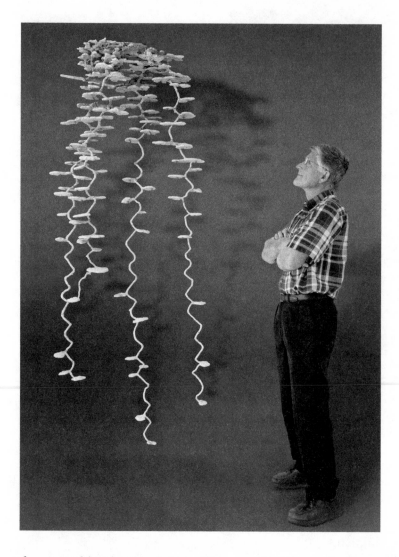

derground kitchen and the millstones that were rolled into place when enemies invaded.

And I was only getting started. I was going to propose a theory to Tschinkel.

Digging, I was going to say, is a primal act—one of the

most basic things humans do. When we dig a hole in the earth and climb down underground, we are engaging in truly eternal behavior, going all the way to the roots of the evolutionary tree, past our earliest mammalian ancestors, past the first vertebrates, down to the origins of multicellular life. As part of the burrowing lineage, I was going to say, we cannot help but feel an old and powerful connection to the earth. Deeper than our fear of enclosure, deeper than our fear of darkness or of being buried alive is a feeling of security we get from digging underground, of feeling the earth embrace us. Perhaps this convergence of ant and human burrows, I was going to say, is a reminder that we are only animals interacting with the earth like any other animal, all of us searching for the same solutions to the same eternal problems.

But before I could even get started, Tschinkel cut me off: "What do you mean by *millstones*?"

"Big circular stones," I said. "Shaped like a donut." I

pulled out my notebook and started to draw a picture for him. "They would roll them into place when—"

Tschinkel was nodding, and when I saw the expression on his face, I stopped.

"There's a species of ant in Costa Rica," he said, "called *Stenamma alas*."

It had just recently been identified by John Longino, a colleague of his at Evergreen State University in Washington. *Stenamma alas*, Tschinkel explained, was perpetually under siege by a particular species of pugnacious army ant. To defend against these attacks, the ant had developed a peculiar adaptation. "They keep a pebble just the right size next to the entrance of the nest," he said. "When army ants invade, the colony retreats into the nest. The last ant in pulls the pebble over the entrance."

Soon after leaving Tallahassee, I emailed John Longino. Enclosed in his reply were photographs of *Stenamma alas* nests, with the pebble next to the entrance. And of an ant, apparently the last of the colony to retreat underground during an invasion, pulling the pebble into place. Longino wrote that he had spoken with Tschinkel the week before and that they had already begun calling *Stenamma alas* the "Cappadocian ant."

CHAPTER 6

THE LOST

Sometimes from sorrow, for no reason,
you sing. For no reason, you accept
the way of being lost, cutting loose
from all else and electing a world
where you go where you want to.

—WILLIAM STAFFORD,
"CUTTING LOOSE"

On the evening of December 18, 2004, in the hamlet of Madiran, in southwestern France, a man named Jean-Luc Josuat-Verges wandered into the tunnels of an abandoned mushroom farm and got lost. Josuat-Verges, who was forty-eight and employed as a caretaker at a local health center, had been depressed. Leaving his wife and fourteen-year-old son at home, he'd driven up into the hills with a bottle of whiskey and a pocketful of sleeping pills. After steering his Land Rover into the large entrance tunnel of the mushroom farm, he'd clicked on his flashlight and stumbled into the dark. The tunnels, which had been originally dug out of the limestone hills as a chalk mine, comprised a five-mile-long labyrinth of blind corridors, twisting passages, and dead ends. Josuat-Verges walked down one corridor, turned, then turned again. His flashlight battery slowly dimmed, then died; shortly after, as he tromped down one soggy corridor, his shoes were sucked off his feet and swallowed by the mud. Josuat-Verges stumbled barefoot through the maze, groping in pitch-darkness, searching in vain for the exit.

On the afternoon of January 21, 2005, exactly thirty-four days after Josuat-Verges first entered the tunnels, three local teenage boys decided to explore the abandoned mushroom farm. Just a few steps into the dark entrance corridor, they discovered the empty Land Rover, with the driver's door still open. The boys called the police, who promptly dispatched a search team. After ninety minutes, in a chamber just six hundred feet from the entrance, they found Josuat-Verges. He was ghostly pale, thin as a skeleton, and had grown out a long, scraggly beard—but he was alive.

In the following days, as the story of Josuat-Verges's survival reached the media, he became known as *"le miraculé des ténèbres,"* "the miracle of darkness."

He regaled reporters with stories from his weeks in the mushroom farm, which seemed to rival even the grandest tales of stranded mountain climbers or shipwreck victims on desert islands. He ate clay and rotten wood, which he found by crawling on all fours and pawing at the mud; he drank water that dripped from the limestone ceiling, sometimes even sucking water from the walls. When he slept, he wrapped himself in old plastic tarpaulins left behind by the mushroom farmers. The part of Josuat-Verges's story that confounded reporters was that he had undergone radical and unexpected oscillations in his mood.

At times, as one might expect, he sank into profound despair; from a piece of rope he found, he even made a noose, "in case things got unbearable." But there were other moments, Josuat-Verges explained, where, as he walked in the dark, he would slip into a kind of meditative calm, allowing his thoughts to soften and unspool, as he embraced the feelings of disorientation, letting himself float through the tunnels in a peaceful detachment. For hours at a time, as he wandered the maze, he said, "I sang to myself in the dark."

When I first read the story of Jean-Luc Josuat-Verges, and his mysteriously ambivalent experience of lostness, my thoughts turned to an ill-conceived excursion I'd made several years before in Paris. Along with two friends—Séléna and Åsa—I decided to go down into the catacombs in order to retrace the path of an eighteenth-century man who, as it happened, had famously entered the quarries and become lost. In 1793, Philibert Aspairt—a man in his sixties, who worked as the guard at the Val-de-Grâce hospital—had descended underground on a quest to find the cellar of a nearby convent, which was said to contain a secret cache of very fine chartreuse. Aspairt lost his bearings, and eleven years later his corpse was found in an alcove beneath the boulevard Saint-Michel. A memorial tombstone was installed on the spot where he fell.

IN THE MEMORY OF PHILIBERT ASPAIRT
LOST IN THIS QUARRY ON NOVEMBER 3, 1793
FOUND ELEVEN YEARS LATER AND
BURIED AT THE SAME PLACE ON APRIL 30, 1804

On a bracingly cold December night, as Séléna, Åsa, and I crouched near the entrance of the catacombs and prepared to crawl underground, I explained that the cataphiles had adopted Philibert Aspairt as their patron saint. On any trip into the quarries, it was customary to pay a visit to the tomb of Philibert, where the cataphiles would leave flowers, votive candles, glasses of wine, even small artworks. We planned to do as the natives did: we would hike to the tomb of Philibert, then retrace our steps and return to the surface a few hours later. The next morning, Séléna and Åsa were due in class, where, as it happened, they were both studying to be professional clowns. At around eight in the evening, equipped with a small bag containing supplies for a short trip—a bottle of wine, a loaf of bread, and a bottle of water—we wriggled through the entrance foxhole and dropped into the two-hundred-mile-wide maze.

I was the navigator, but, in retrospect, an entirely incompetent one. At the time, I'd only just arrived in Paris—this was years before the traverse of the city with Steve Duncan—and had visited the catacombs only once before. I had a copy of a map of the quarries, which I'd obtained from an explorer named Hatchet, but had not actually employed it in navigation. Hatchet had briefly pointed out the location of the major landmarks in the quarries, as well as the entrance, which was not specifically marked on the map. It was a cursory tutorial, and I should have asked more questions, but at the time I'd assumed I would be able to orient myself once I got underground.

We turned in the dark, then turned again, weaving through the rocky honeycomb, our headlamp beams glancing down the walls, water splashing at our boots. It was Séléna's and Åsa's first time down in the quarries: they listened to the far-off whisper of the metro and ran their hands over the cool stone. We had been walking for perhaps an hour when we entered a tight, low-ceilinged chamber where the mud on the floor had dried, leaving a pattern of cracks etched at our feet. Crouching over the floor, I commented that the cracks resembled the passages of a maze, as though we were peering inside a microcosmic model of the network we were at that moment navigating, a maze nested inside of a maze.

Which is about when I recognized my error. I was looking at the map, trying to figure out the next turn in our path to the tomb of Philibert, when I realized, with a sudden pitch in my stomach, that I'd been wrong about the location of the entrance on the map. Which is to say, every turn we'd made from the moment we went underground had been a wrong turn. We were nowhere near the tomb of Philibert—I had no idea where we were. I had not the faintest clue how far we'd gone, or how to get back, or even which direction we were facing. I explained to Séléna and Åsa, in a shrinking voice, what had happened. We all fell silent. We had limited food and water. We had dwindling batteries in our headlamps. We had no compass.

———

HOMO SAPIENS HAVE ALWAYS been marvelous navigators. We possess a powerful organ in the primitive region of our brain called the hippocampus, where, every time we take a step, a million neurons collect data on our location, compiling what neuroscientists call a "cognitive map," which keeps us always oriented in space. This robust apparatus, which far outstrips our modern needs, is a hand-me-down from our nomadic hunter-gatherer ancestors, whose very survival depended on powers of navigation. For hundreds of thousands of years, the failure to locate a watering hole or a safe rock shelter, or to follow herds of game and locate edible plants, would lead to certain death. Without the ability to pilot ourselves through unfamiliar landscapes, our species would not have survived—it is intrinsic to our humanity.

It is no surprise, then, that when we *do* lose our bearings, we are cast into a primal, bitter-in-the-mouth panic. Many of our most elementary fears—being separated from loved ones, uprooted from home, left out in the dark—are permutations of the dread of being lost. In our fairy tales, it is when the fair maiden becomes disoriented in the gloomy forest that she is accosted by the menacing troll or the hooded crone. Even hell is often depicted as a maze, going back to Milton, who made the comparison in *Paradise Lost*. The archetypal horror story of disorientation, of course, is the Greek myth of the Minotaur, who dwells in the winding folds of the Labyrinth of Knossos, a structure, as Ovid wrote, "built to disseminate uncertainty," to leave the visitor "without a point of reference."

So deep-seated is our dread of disorientation that becoming lost may trigger a kind of crack-up, where our very sense of self comes apart at the seams. "To a man totally unaccustomed to it," wrote Theodore Roosevelt in his 1888 book *Ranch Life and the Hunting Trail*, "the feeling of being lost in the wilderness seems to drive him into a state of panic terror that is frightful to behold, and that in the end renders him bereft of reason. . . . If not found in three or four days, he is very apt to become crazy; he will then flee from the rescuers, and must be pursued and captured as if he were a wild animal."

We may get lost roaming the blank Arctic tundra, or hacking through dense jungle—but the ultimate arena of lostness is the underground world. To lose your bearings in a warren of subterranean hollows is its own species of disorientation. In a cave, as Mark Twain wrote of McDougal's Cave, where Tom Sawyer and Becky Thatcher became lost for three days, "one might wander days and nights together through its intricate tangle of rifts and chasms, and never find the end . . . he might go down, and down, and still down, into the earth, and it was just the same—labyrinth under labyrinth, and no end to any of them." From our first step into subterranean darkness, our hippocampus, which so reliably guides us through the surface world, goes on the fritz, like a radio that has lost reception. We are cut off from the guidance of the stars, from the sun and the moon. Even the horizon vanishes—if not for gravity, we'd scarcely know up from down. All of the subtle cues that may orient us on the surface—cloud formations, plant growth patterns, animal tracks, wind direction—disappear. Underground, we lose even the guide of our own shadow.

When we climb a mountain, or push out to sea, we recede from familiar territory: we may peer behind us to see how far we've traveled, and squint ahead to see what lies before us. Down in a tight cave passage, or in the bounded folds of a catacomb, our field of view is blinkered, never reaching beyond the next twist or kink. As the cave historian William White observed, you never really *see* a whole cave—only one sliver at a time. When we navigate a landscape, wrote Rebecca Solnit in *A Field Guide to Getting Lost*, we are reading our surroundings as a text, studying "the language of the earth itself"; the underground is a blank page, or a page scribbled with language we cannot decipher.

Not that it's illegible to everyone: certain subterranean-dwelling creatures are marvelously adapted to navigate through the dark. We all know the bat, who swoops through cave darkness using sonar and echolocation, but the champion subterranean navigator may be the blind mole rat: a pink, wrinkly, bucktoothed creature—picture a ninety-year-old thumb with fangs—that spends its days in vast, mazelike underground nests. To navigate these dark passages, the blind mole rat periodically drums its head against the ground then discerns the shape of the space according to the patterns of the returning vibrations. In its brain, the rat even has a tiny iron deposit, a built-in compass, which detects the earth's magnetic field. Natural selection has endowed us surface-dwellers with no such adaptive tricks. For us, a step underground is always a step into a navigational vacuum, a step in the wrong direction, or rather, no direction at all.

DOWN IN THE CATACOMBS, I mumbled my way through apology after apology, but Séléna and Åsa hushed me. There was no point, they said, in wasting our energy on panicking. (To be fair, Séléna cast me a look that left open a door for future admonishment.) For now, our goal was plain. The only way out of the maze was the hole in the wall through which we'd entered: we had to retrace our steps.

Séléna and Åsa, whose years of improvisation in clown troupes had endowed them with superlative skills of team communication, orchestrated a precise and democratic plan. We would move back through the tunnels systematically, searching for pronounced oddities in the rock, memorable graffiti, or conspicuous footprints in the mud. Starting at each intersection, we would explore every possible tunnel one by one: when we were positive we recognized nothing, we would backtrack and explore the next branch. We would follow a tunnel only when we all agreed that it was familiar.

It proved to be a vexing, exhausting process. Every passage resembled every other, all with the same stony contours, all tumbling out in erratic geometry. Much of the graffiti at each intersection had been painted by past visitors as trail markers—arrows, stars, and geometric shapes—to help them keep track of their route back to the entrance. We attempted to isolate one thread and follow it, imagining that perhaps the blue triangles or red circles would lead us to the exit. But we soon gave up, as all of the trail markers blended together in a noisy clutter. It was like wandering into a folktale forest to discover hundreds of bread-crumb trails snaking among the trees.

Down one tunnel, we thought we heard the soft hiss of a cataphile's carbide lamp coming from an adjacent gallery, but when we called out, no one answered. Around another corner, we discovered a stone staircase spiraling up into the dark: Séléna and I ascended until we found ourselves crouching directly beneath a manhole lid leading to the street. I tried to push it ajar with my shoulder, but I was either not strong enough or it had been sealed shut. So close to the surface, it occurred to us that we might have enough cellphone service to call for help, but as I puzzled over whom I might call, how I might explain our situation, and what on earth I might suggest they do to help, we realized that all three of our phones had died. Åsa, meanwhile, did her best to keep the mood light. When we circled back to an intersection that we all knew we'd passed seven or eight times, she would stop and say, "Guys!" Séléna and I would turn back, and Åsa, wide-eyed, would whisper, "I think we've been here before."

We all maintained an outward calm, but as the hours passed, and we continued fumbling in the dark, we began to make unsettling calculations in our heads. We started rationing battery power in our headlamps, one of us lighting the way, while the other two switched their lights off. During breaks, we grimly eyed the level of water in the bottle, taking only sparing sips. We refrained altogether from eating the bread, imagining that we might require it for energy in the hours ahead. Each time we lost our thread in the tunnels and needed to reset, we retreated to a small, cramped chamber that happened to house a cataphile sculpture: a plaster figure of a man built half inside the limestone wall, as though he were trapped inside the rock.

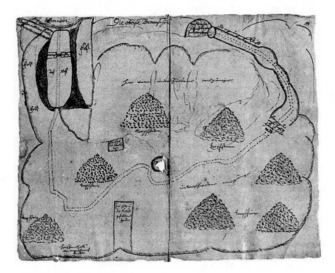

IN ANY OTHER LANDSCAPE, when our inborn powers of navigation falter, we turn to a map, which anchors us in space, and keeps us on course. In the underground world, though, mapping has always been a uniquely perplexing endeavor. Long after explorers and cartographers were charting every other terrestrial landscape on the planet, casting clean latitudinal and longitudinal grid lines over remote archipelagos and mountain ranges, the spaces directly beneath our feet remained elusive.

The earliest known map of a cave was drawn in 1665 of Baumann's Cave, a large cavern situated in the densely forested Harz region of Germany. To judge from the map's rudimentary lines, the cartographer, a man identified as Von Alvensleben, does not appear to have been an expert mapmaker, or even a capable one, but the map's shortcomings are nonetheless remarkable. The explorer has failed to convey any sense of perspective, or depth, or any other dimension—he has failed to communicate even that the space is underground. Von Alvensleben was attempting to map a space he was neurologically ill-equipped to see, a space literally beyond his perception. It came to the point of an epistemological folly, like trying to paint a portrait of a ghost, or catch a cloud in a net.

The map of Baumann's Cave was the first in a long lineage of curious failures of subterranean cartography. For generations, explorers all over Europe—teams of dauntless, quixotic men—plumbed caves with the intent to measure the underground world, to orient themselves in the dark, only to fail, often in bewildering ways. On fraying ropes, they lowered themselves deep underground, where they wandered for hours, clambering over hulking boulders and swimming down subterranean rivers. They guided their way with wax candles, which gave off feeble coronas of light that extended no more than a few feet in any direction. One seventeenth-century explorer descended into a cave in England, where he attempted to measure the dimensions of one chamber, only to realize that he couldn't even *see* the boundaries of the space, never mind measure them: "By the light of candles," he wrote, "we could not fully discern the roof, floor or sides of it." Surveyors often resorted to absurd measures, such as an Austrian explorer named Joseph Nagel, who, in an attempt to illuminate a cave chamber, tied a rig of candles to the feet of two geese, then threw pebbles at the geese, hoping that they would take flight and cast their light through the dark. (It didn't work: the geese wobbled lamely and tumbled earthward.) Even when they did manage to make measurements, meanwhile, the explorers' spatial perception was so warped by the caprices of the environment that their findings would be wildly off the mark. On a 1672 expedition in Slovenia, for example, an explorer plumbed a winding cave passage and recorded its length at six miles, when in reality, he had traveled only a quarter mile. The surveys and maps that emerged from these early expeditions were often so divergent from reality that some caves are now effectively unrecognizable. Today, we can only read the old reports as small, mysterious poems about imaginary places.

The most renowned of the early cave mappers was a late-nineteenth-century Frenchman named Edouard-Alfred Martel, who would become known as the "father of speleology." Over the course of a five-decades-long career, Martel led some 1,500 expeditions in fifteen countries around the world, hundreds of them into virgin caves. A lawyer by trade, he spent his early years rappelling underground in shirtsleeves and a bowler cap, before finally designing a kit of specialized caving equipment. In addition to a collapsible canvas boat dubbed *Alligator*, and a chunky

field telephone to communicate with porters on the surface, he devised a battery of subterranean survey instruments. For example, he invented a contraption to measure
a cave floor-to-ceiling, in which he attached an alcohol-
soaked sponge to a paper balloon on a long string, then set
a match to the sponge, causing the balloon to rise to the
roof as he unspooled the string. Martel's maps may have
been more precise than those of his predecessors, but compared to the maps drafted by explorers of any other landscape at the time, they were hardly more than sketches.
Martel was celebrated for his cartographic innovation of
dividing a cave into distinct cross sections (or *coupes*), which
would become the standard in cave mapping. To my eye,
though, the maps only feel like further testament to the
slipperiness of the subterranean environment, as though
each map were a record of what could not be mapped. The
only way to fully comprehend an underground space, they

tell us, is to break it down into fragments and lay them out like the bones of a disarticulated skeleton.

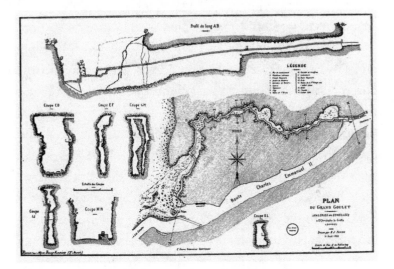

Martel and his fellow explorers, who spent years trying and failing to orient themselves in the subterranean world, were disciples to lostness. No one knew the sensory experience of disorientation so intimately: for hours on end, they'd float through the dark, caught in a prolonged state of vertigo, as they tried and failed to anchor themselves. According to all evolutionary logic, where our minds are wired to avoid disorientation at all costs, where lostness activates our most primitive fear receptors, they must have experienced a deep anxiety: "the panic terror that is frightful to behold," as Roosevelt described it. And yet, they went down again and again. They lowered themselves into uncharted hollows, where no human had set foot before them, where the residents of local villages would not even peer over a rocky edge for fear of spirits dwelling below.

They derived a form of power, it seems, from losing themselves in the dark.

In 1889, Martel mounted a campaign to explore a giant cave in southwestern France called the Gouffre de Padirac. On a July afternoon, he and his team strapped into ropes and lowered themselves two hundred feet down into the chasm, slowly descending past clumps of soft green vegetation growing in the ridges of the walls. When they touched down on the rocky bottom, where the air was cool and damp, the stones coated in moss, they found a subterranean river vanishing into a crevice in the wall. They lit their candles, boarded the *Alligator*, and slowly paddled into the dark, where they floated beneath soaring galleries, ducking around heaving curtains of stalactites. The water dripping around them in the dark sent up "a melodious song," as Martel wrote, "sweeter and more harmonious than the dulcet tones of the upper world." They followed one branch of the river, then another, until they felt themselves utterly cut off from all known geography. For twenty-three hours, they drifted in a perfect vacuum.

"The unknown draws us irresistibly forward!" Martel wrote of the expedition. "No man has gone before us in these depths. No one knows where we go nor what we see, nothing so strangely beautiful was ever presented to us, and spontaneously we ask each other the same question: are we not dreaming?" In Martel's words, I hear a man lifted into a strange kind of rapture. When I picture the explorer fumbling through the deep folds of Padirac, holding the flame of his waxen candle overhead, I can almost hear him softly singing to himself in the dark; the same song, perhaps, that

Jean-Luc Josuat-Verges sang as he wandered the tunnels of the mushroom farm in Madiran.

LOSTNESS HAS ALWAYS BEEN an enigmatic and many-sided state, always riven with unexpected potencies. Across history, all varieties of artists, philosophers, and scientists have celebrated disorientation as an engine of discovery and creativity, both in the sense of straying from a physical path, but also in swerving away from the familiar, turning in to the unknown. "Not to find one's way around a city does not mean much," wrote Walter Benjamin. "But to lose one's way in a city—as one loses one's way in a forest—that requires quite a different schooling." To make great art, John Keats said, one must embrace disorientation and turn away from certainty. He called this "Negative Capability": "that is, when a man is capable of being in uncertainties, mysteries, doubts, without any irritable reaching after fact and reason." Thoreau, too, described lostness as a door into understanding your place in the world: "Not till we are completely lost, or turned round," he wrote, "do we appreciate the vastness and strangeness of nature. . . . Not till we are lost, in other words, not till we have lost the world, do we begin to find ourselves, and realize where we are and the infinite extent of our relations." For Solnit, meanwhile, lostness is the ultimate way to "become fully present" to our surroundings. "One does not get lost," she wrote, "but loses oneself, with the implication that it is a conscious choice, a chosen surrender, a psychic state achievable through geography."

All of which makes sense, neurologically speaking: when we are lost, after all, our brain is at its most open and absorbent. In a state of disorientation, the neurons in our hippocampus are frantically sponging up every sound, smell, and sight in our environment, scrambling for any strand of data that will help us regain our bearings. Even as we feel anxious, our imagination becomes prodigiously active, conjuring ornate images from our environment. When we take a wrong turn in the woods and lose sight of the trail, our mind perceives every twig snap or leaf rustle as the arrival of an ornery black bear, or a pack of warthogs, or a convict on the lam. Just as our pupils dilate on a dark night, in order to receive more photons of light, when we are lost our mind opens up to the world more fully.

———

ONCE, WHILE EXPLORING BENEATH the streets of Naples, I experienced a brief, transcendent lostness. On a fall morning, near the ancient center of the city, two urban explorers—Luca and Dani, brothers—guided me down into the basement of an ancient basilica, where we found an enormous hole in the floor. After clipping into ropes, we climbed down a long ladder, then rappelled into the basin of an ancient Greek cistern, a large bottle-shaped hollow, nearly a hundred feet beneath the city. The cistern, Luca and Dani explained, was but a single node in a maze of hollows that had been expanding beneath Naples since the eighth century B.C. The network comprised a tangle of crypts, catacombs, tombs, and cisterns so vast and labyrinthine, that no one truly knew its limits.

As morning turned to afternoon, we climbed from one cistern to the next, following narrow sidewinding tunnels deeper and deeper into the maze. Every hollow began to resemble every other, until it felt like we were moving through a three-dimensional hall of mirrors, like a set piece from a Borges story. Some cisterns branched into multiple channels, and those channels branched again, radiating out in wild fractal patterns.

We soon broke into unknown parts of the network, dropping into chambers that Luca and Dani had not seen previously; with each discovery, they would shout in the dark, in the way of far-flung mariners spotting uncharted islands on the horizon. And then at one point, without my even noticing—an event that seemed to move directly from future to past without ever being in the present—I was alone in the dark. One moment, we'd all been together in a cistern, then I turned to set up a photograph, and when I turned back around, Luca and Dani were gone. I crawled down the passageway I thought they'd taken, but ended up in an empty chamber. Suddenly the metallic clink of carabiners against my hip was the only sound I could hear, and I could see nothing beyond the cone of my headlamp beam. I called out for them, but heard no response, my voice fading in the winding corridors.

We were not separated for long—no more than a few minutes. But even in that brief period, as it dawned on me that I had no idea how far I'd traveled, no clue where I was relative to any previous point, I experienced a complete and utter unmooring. It was the feeling of being cut loose, as though my feet had lifted off the ground and I was falling through space. What I felt was not exactly panic, but a searing clarity, an almost amphetaminic alertness, where all of my senses prickled awake, and I felt myself entirely immersed in the present, attuned to the faintest smells, sounds, and agitations of air that had previously eluded me. Even my skin became alert, as though I were absorbing the world through my pores.

"GETTING LOST," WROTE SOLNIT, "is like the beginning of finding our way, or finding another way." When we stray from our course, and our nerves turn raw, our relationship to the world becomes malleable. Even our most deeply ingrained beliefs and thought lines may come undone as we open ourselves to new interpretations of reality. In the literature of religion, it is when we are lost that we receive the burst of revelation, undergo the conversion, or experience the mystical awakening. In the Old Testament, the prophets are lost in the desert just before they find God. Siddhartha Gautama spends six years wandering, untethered from fixed geography, before he becomes the Buddha. Dante's spiritual quest in the *Inferno*, meanwhile, begins with a declaration of lostness: "In the middle of the journey of our life / I came to myself, in a dark wood, where the direct way was lost." The novelist Jim Harrison once told the poet Gary Snyder that when we are lost, "Suddenly everything is in question, including your own nature. It's that dramatic. . . . I've often thought that being lost is like a *sesshin* [a period of meditation] when you sit for a long time and then a gong goes off and you get up and the world looks completely different." To which Snyder responded, "Well, that is like enlightenment."

In the late 1990s, a team of neuroscientists tracked the power of disorientation down into the physical trappings of our brain. In a lab at the University of Pennsylvania, they conducted experiments on Buddhist monks and Franciscan nuns, where they scanned their brains during the act of meditation and prayer. Immediately, they noticed a pattern: in a state of prayer, there was a small region near the front of the brain, the posterior superior parietal lobe, that showed a decline in activity. This particular lobe, as it turns out, works closely with the hippocampus in the processes of cognitive navigation. As far as the researchers could see, the experience of spiritual communion was intrinsically accompanied by the dulling of spatial perception.

It should be no surprise, then, that anthropologists have tracked a kind of cult of lostness running through the world's religious rituals. The British scholar Victor Turner observed that any sacred rite of initiation proceeds in three stages: separation (the initiate departs from society, leaving behind his or her former social status); transition (the initiate is in the midst of passing from one status to the next); incorporation (the initiate returns to society with a new status). The pivot occurs in the middle phase, which Turner called the "stage of liminality," from the Latin *limin,* meaning "threshold." In the liminal state, "the very structure of society is temporarily suspended": we float in ambiguity and evanescence, where we are neither one identity nor the other, no-longer-but-not-yet. The ultimate catalyst of liminality, Turner writes, is disorientation.

Among many rituals of lostness practiced by cultures all over the world, a particularly poignant one is observed by the Pit River Native Americans in California, where, from time to time, a member of the tribe will "go wandering." According to anthropologist Jaime de Angulo, "the Wanderer, man or woman, shuns camps and villages, remains in wild, lonely places, on the tops of mountains, in the bottoms of canyons." In the act of surrendering to disorientation, the tribe says, the Wanderer has "lost his shadow." It is a mercurial endeavor to go wandering, a practice that may result in irredeemable despair, or even madness, but may also bring great power, as the Wanderer emerges from lostness with a holy calling, before returning to the tribe as a shaman.

But the most ubiquitous vehicle of ritual lostness—the most basic embodiment of disorientation—is the labyrinth. We find labyrinthine structures in every corner of the world, from the hills of Wales to the islands of eastern Russia to the fields of southern India. A labyrinth operates as a kind of liminality machine, a structure devised to engineer a concentrated experience of disorientation. As we enter the winding stone passages, and turn our focus to the bounded path, we disconnect from external geography, slipping into a kind of spatial hypnosis, where all reference points fall away. In this state, we are primed to undergo a transformation, where we pass between social statuses, phases of life, or psychic states. In Afghanistan, for example, labyrinths were the center of marriage rituals, where a couple would solidify their union in the act of navigating the twisting stone path. Labyrinthine structures in southeast Asia, meanwhile, were used as meditation tools, where a visitor would walk slowly along the trail in order to deepen their inward focus. Indeed, the archetypal tale of Theseus slaying the Minotaur in Crete is ultimately a story of transformation: Theseus enters the labyrinth as a boy and emerges a man and a hero.

In their modern incarnation, most labyrinths are two-dimensional, their passages bordered by low stacks of stones or mosaic patterns tiled into a floor. But as we trace the lineage of the labyrinth deeper into the past, searching for earlier and earlier incarnations, we find the walls slowly rising, the passageways becoming darker and more immersive—indeed, the very first labyrinths were almost always underground structures. The ancient Egyptians, according to Herodotus, built a vast subterranean labyrinth, as did the Etruscans in northern Italy. The pre-Incan culture of Chavín constructed an enormous underground labyrinth high in the Peruvian Andes, where they conducted sacred rituals in dark, sinuous tunnels; the ancient Maya did the same in a dark labyrinth in the city of Oxkintok in the Yucatán. In the Sonoran Desert of Arizona, meanwhile, the Tohono O'odham tribe have long worshipped a god called I'itoi, also known as the Man in the Maze, who dwells at the heart of a labyrinth. The opening of I'itoi's labyrinth, a design frequently woven into the tribe's traditional baskets, is said to be the mouth of a cave.

The very first depiction of a labyrinth anywhere in the world, according to a 1998 archaeological study in northwest Sicily, is a five-thousand-year-old painting, which was discovered deep in the dark zone of a cave. Archaeologists have hypothesized that there was once a labyrinth laid out on the cave's muddy floor at the foot of the painting, which functioned as the ceremonial path in an ancient rite of passage. A plausible explanation, to be sure. But I tend to wonder if perhaps the cave itself was the labyrinth. I wonder if the painting, rather than referencing a separate structure, was illustrating the sensation of entering the cave, of losing yourself in the dark and drifting through stony passageways.

When Jean-Luc Josuat-Verges entered the tunnels of

the mushroom farm in Madiran with his whiskey and sleeping pills, he'd had notions of suicide. "I was low, having very dark thoughts" was the way he put it. After he emerged from the maze, he found that he'd regained his purchase on life. He rejoined his family, where he found himself happier and more at ease. He began attending night school, earned a second degree, and found a better job in a town up the road. When asked about his transformation, he told reporters that while he was in the dark, "a survival instinct" had kicked in, renewing his will to live. In his darkest moment, when he desperately needed to transform his life, he traveled into the dark, surrendered to disorientation, preparing himself to emerge anew.

———

IN THE END, what saved Séléna, Åsa, and me—our Ariadne's thread—was the winter air. The catacombs maintain a year-round temperature of about 57 degrees Fahrenheit, which, on that particular December night, was about 20 degrees warmer than the temperature on the surface. As we fumbled through the tunnels, searching for any recognizable landmark or point of orientation, we began to feel something unexpected: faint whispers of cool air. It would blow gently on our skin, then fade for a moment, then return, then fade. Slowly we pieced together that the air was coming from the exit hole. And so we followed the cold. If we crawled down a tunnel and the air began to feel warmer, we'd turn back, knowing we were headed the wrong direction. Had it been a few months later—say, a mild spring night—when the aboveground-underground temperature differential was less clear, we might never have found our way out.

As we came to the ragged exit hole, the three of us heaved ourselves out of the tunnel, drinking in frigid air. It was past four in the morning—we'd been lost for eight hours. We climbed back up to street level, laughing and whooping down the empty boulevard. With the metro closed, we headed back to Séléna's apartment in a cab, where the driver eyed the three of us in the rearview mirror, soggy and muddy and elated in the backseat. We sat on blankets on the floor of Séléna's tiny studio, beneath an angled skylight, and toasted to our survival. As dawn filtered into the room, we dissected the curiosities of being lost in the dark, replaying the night's events, sharing what had been going through our heads at various moments. Each of us at times had felt a creeping anxiety, each of us had felt dread. But beneath that, in a more obscure channel of the mind, we'd all found moments of lucid calm, where we were briefly lifted outside of ourselves.

THE
HIDDEN BISON

All sacred things must have their place.
It could even be said that being in their place
is what makes them sacred.

—CLAUDE LÉVI-STRAUSS,
The Savage Mind

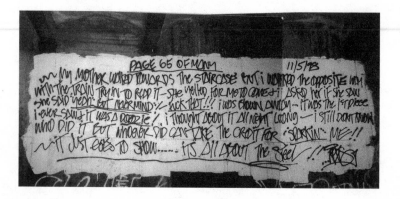

In my earliest days of exploring New York, when I spent afternoons riding the subway, peering hawk-eyed out the window, searching for passageways to abandoned stations,

I would, from time to time, glimpse mysterious messages written on the tunnel walls. They were rectangular panels of white or yellow paint, maybe five feet high by ten feet wide, covered in black lettering. They appeared always in the dark parts of the tunnel, in the empty, soot-covered no-man's-land between stations. Once I started looking, I began to see the panels everywhere, blinking through the frame of a train window as I weaved beneath quiet neighborhoods in Brooklyn, or under buzzing avenues in Midtown Manhattan. I could never actually read the text—as the train rushed past, I'd only ever catch a few letters—but they nonetheless fascinated me, these subliminal messages flickering in the unconscious of the city.

The panels, I eventually learned, were part of a mysterious art project by a graffiti writer called REVS. Each was a "page" in a diary, written over the course of six years, and scattered throughout New York's subterranean corridors. There were 235 pages in all, one painted between almost any two subway platforms in the city. Late at night, REVS would put on a hardhat and fluorescent vest to disguise himself as an employee of the MTA, then he'd drop through an emergency exit hatch in the street. Down in the dark, he'd use a roller to put up a large rectangle of bucket paint on a tunnel wall, then, with a can of black spray paint, he'd write a paragraph-long entry, a childhood vignette or a short philosophical rumination.

REVS, I learned, was a luminary of New York graffiti. In a culture that champions the prolific, where an artist strives to paint his or her name on the face of the city as many times as possible, in the most visible places without ever getting caught, no one was so omnipresent as REVS. Going back to the early 1980s, he'd left his mark on the city tens of thousands of times. With spray paint and markers, he tagged the sides of telephone booths, newspaper dispensers, and mailboxes; he painted billboard-sized murals on brick façades; he fastened canvases to the sides of buildings with bolts and screws; he even welded metal sculptures of his tag to street signs and iron fences. During the late 1980s and early 1990s, when REVS was at his most prolific, it was hard for New Yorkers to walk more than a few steps on any city block without seeing the four letters of his name, as though he were whispering a quiet song in the ears of the city. Around that time, when Mayor Rudolph Giuliani enlisted a cell of the MTA called the Vandal Squad to scrub the city clean of graffiti, REVS became Public Enemy Number One. The Kingfish, they called him. As his notoriety grew in the streets, REVS began going underground, where he started painting his life story in the dark.

In an old history book on New York graffiti, I found images of the earliest entries in the diary, which had been painted on the walls of subway tunnels beneath Brooklyn. On a page dated March 5, 1995, the story began with the birth of the author.

DEAR SOCIETY,

I WAS BORN ON APRIL 17, 1967 IN BROOKLYN, NY. THE HOSPITAL WAS VICTORY MEMORIAL IN BAY RIDGE—I AM AN ONLY SON EXCEPT FOR A HALF BROTHER FROM MY FATHERS 1ST MARRIAGE HIS

NAME IS SEAN AND HE'S AN EX-CON. HE'S A REAL
ASSHOLE CAUSE HE STOLE 2100 DOLLARS FROM
MY UNCLE PATTY WHO TRIED TO HOOK HIM UP
AND GIVE HIM A BREAK - ANYWAY FUCK HIM!!! SO
IT WAS 3 PM ON A MONDAY . . . WEIGHED IN AT 8
LBS 13 OZ BUT SHE HAD TO HAVE A C-SECTION
TO GET ME OUT!! GOOD THINGS NEVER ARE
EASY!! TO BE CONTINUED . . .

Before that was a prologue, marked "1 of Many":

TO JOE PUBLIC . . . YOU MIGHT BE ASKIN' YOUR-
SELF RIGHT NOW, WHAT IS THIS . . . IT'S ABOUT
A KID WHO IS JUST LIVING HIS LIFE AND TELLIN'
HIS STORY THE ONLY WAY HE KNOWS HOW.
THE THING IS HIS STORY AIN'T NO DIFFERENT

THAN YOURS—WE ALL END UP BEING PIECES
TO A PUZZLE FROM WHICH THERE IS ONLY
ONE ALMIGHTY CREATOR. GOD

I wondered. What kind of person only knows how to tell his story by writing it on subway tunnel walls? Who was REVS?

All graffiti writers live cloaked lives, but no one, I learned, was quite so invisible as REVS. The younger graffiti writers revered REVS, but had never met him; older writers knew him, but had not seen him in years. The few writers who were in contact with REVS, meanwhile, scoffed when I asked if they could put us in touch. "Not a chance," said a writer called ESPO. "REVS doesn't talk to anybody," SMITH told me. A photographer who'd once taken a portrait of REVS (with his face obscured) told me, "Even if I knew where he was, which I don't, I wouldn't tell you." For years, REVS drifted in and out of my thoughts. I'd forget about him for a few months at a time, but then I'd glimpse a page streaking past a subway window and I'd imagine a shrouded man painting down in the dark, and become newly fixated on the search.

I crisscrossed the city, following even the faintest trails. I walked the street where I was told he'd once kept a welding studio, and roamed his old neighborhood of Bay Ridge, where I questioned the owner of a candy store I'd heard REVS had frequented as a kid. Following a tip that he'd worked as a bridge builder, I called up steel-working unions. I sought the help of officer Steve Mona, who as head of the Vandal Squad had spent nearly a decade chasing REVS.

Mona was no help, nor was anyone else. A photographer who'd been following REVS's work for twenty years told me to give up the pursuit: "You're going to drive yourself crazy. The man is a phantom." And eventually, I did give up, accepting the fact that REVS would remain in the shadows. It was foolish, I told myself, to expect a man who wrote in the dark to step out from behind the curtain and introduce himself. The only way to learn about the author would be to go and read his diary underground.

On a stifling summer night, Russell and I waited on a subway platform for a train to pass, and for the last passenger to disappear through the turnstile, then walked briskly to the edge of the platform, stepped over the DO NOT ENTER OR CROSS TRACKS gate, and headed into the dark. The air was stale and heavy, and the tunnel rippled with the sounds of water dripping from the ceiling. We had not been walking long before we found his trail. As our eyes began to adjust in the dark, we saw tags—R-E-V-S—painted vertically on the iron girders separating the express and local tracks. And then, on the other side of the girders, we found ourselves reading a diary page. Faded and smudged with steel dust, it looked like it had been hanging on the wall for centuries. It was a description of two brothers from the old neighborhood—Chris and Danny—with whom REVS used to listen to KISS and watch *Saturday Night Live*. All told, a pretty unremarkable story. And yet, down in the dark of the tunnels, with the city far above our heads, we read it breathlessly, as though we'd uncovered some ancient runic poem.

Some trips to read the diary were impromptu, others

planned. I'd go alone, or with Russell, or with other friends who'd grown tired of hearing me talk about REVS, and wanted to see a page for themselves. Steve and other urban explorers talked about finding a kind of tranquility in running the subway tunnels, but in my experience, from the moment I stepped off the edge of one platform until the moment I was safely standing on the next, I was racked with anxiety. The curved parts of tunnels frightened me, because trains came around with less warning. Unnerving, too, were narrow stretches that offered no clearance: they were marked with red-and-white-striped panels, which were known to graffiti writers as "blood and bones." I'd get jumpy on rainy nights, because I couldn't distinguish the sound of raindrops from the clinking sounds of an oncoming train. The sight of a camera angled at the entrance of a tunnel always spooked me, and sent me hightailing back to the platform. (Steve told me that a camera meant no one was paying attention, but I never quite trusted that theory.)

There was the third rail, and the rats, and the threat of being caught by a track worker, and always the specter of the passenger-less black-and-yellow utility trains, which followed no predictable schedule, and could whip around a corner without warning. But on each trip underground, halfway between the platforms, at the darkest point of the tunnel, I'd encounter a page from REVS's diary, and it never failed to leave me awed, as I studied it for telltale clues about the author.

I cherished the diary. I kept a careful transcription of the pages I encountered, as well as those I saw in photographs by other explorers. I read about REVS as a boy causing mischief in Bay Ridge ("Every mother fuckin day it was stickball, whiffleball, stoopball, off the wall, football, skully, kings, skateboarding, manhunt, climbin roofs or fire escapes always battlin"); REVS fighting with his alcoholic father ("On several occasions the cops or an ambulance would bring him home and I would even see him passed out on the sidewalk"); REVS seeing his first fully painted subway car ("I was blown away—it was the 1st piece I ever saw and it was a doozie! I thought about it all night long— I still don't know who did it but whoever did can take the credit for sparkin' me!"). In the way of all New York transplants, I carried nostalgia for how the city used to be, for the New York I'd loved in books and movies, but which had vanished by the time I arrived. REVS's diary was an artifact from that irretrievable New York: a long, gritty punk rock poem from a bygone time.

But above all, what fascinated me about the diary, what left me baffled, was *where* REVS wrote. On one trip under

Brooklyn, I was crouching in an emergency exit alcove, my face streaked in steel dust, my T-shirt damp with sweat and loose around my neck, when I spotted a page on the opposite wall. It was a short entry about an older graffiti writer named ENO, who had been a kind of graffiti sensei for REVS when he was coming up in the ranks. A lamp fixed above the tracks cast a halo of light over the concrete wall: other graffiti writers had tagged within the radius, so that passengers on trains would glimpse their names through the window. Not REVS. He'd placed the page several feet beyond the edges of the glow, in the dark, where it was all but invisible. It was like a poet inscribing sonnets in invisible ink, or a composer writing symphonies in subsonic tones. Why make an artwork, I wondered, in a hidden place, in the darkest part of the city, where almost no one could see it?

IT WAS REVS that I had in mind one sun-streaked November morning as I followed a narrow, twisting road into the Pyrenees in the southwest of France. The Pyrenees, which span the border of France and Spain, from the Bay of Biscay to the Mediterranean Sea, are positively wormeaten with caves. Over the last century and a half, archaeologists have learned that the caves here conceal ancient artworks made by hunter-gatherer tribes who roamed the region between roughly fifty thousand and eleven thousand years ago. The first major discovery came in 1879, when a Spanish nobleman named Marcelino Sanz de Sautuola and his eight-year-old daughter, Maria, went exploring in a cave on his property in the Basque country. Sautuola was scratching at the floor of the cave, when Maria looked up at the ceiling, where, incredibly, she saw a painted herd of golden-red bison. During the following years, in the limestone hills of western Europe, a story became familiar: a farmer, or a shepherd, or a group of local boys, would come upon a fissure in a stony hillside, climb into the dark with a candle, and find themselves blinking in the half-light at ancient images of mammoths, bison, aurochs, ibex, and horses, all executed with stunning vitality and grace.

These artworks, which have now been discovered in 350 caves across Europe, have puzzled archaeologists and anthropologists for generations. At the beginning of the twentieth century, researchers suggested that they were made to fill leisure hours, that Paleolithic people were "making art for art's sake." Subsequent researchers hypothesized that they were remnants of "sympathetic hunting magic," meaning the artists were painting images of their prey in order to cast a spell and make them easier to catch. Today, most archaeologists suspect that the paintings were connected to some form of religious ritual. But the outstanding mystery, which has confounded researchers from the beginning, was that the artworks were placed always in hidden places, in the hardest-to-reach nooks, at the very backs of caves.

I'd come to the Pyrenees to visit what may be the most spectacular, most bewildering, and the most deeply hidden artwork anywhere in the world: *les bisons d'argile,* "the clay bison." They were a pair of clay sculptures, placed more than a half mile inside a large cave, at the end of long, tortuously narrow passageways, in the deepest, most inaccessible chamber. They were made fourteen thousand years ago, by a culture known to archaeologists as the Magdalenians. The cave, which was called Le Tuc d'Audoubert, was one in a system of three caves formed by the Volp River, outside the hamlet of Montesquieu-Avantès in Ariège. Le Tuc was located on the estate of Pujol, which was owned by the Bégouëns, an aristocratic family who'd lived in the region for many generations. The current lord of the chateau, and the devoted custodian of the caves, was Count Robert Bégouën.

That I'd been invited to visit the cave at all was somewhat miraculous. Every archaeologist I'd contacted about Le Tuc had told me that it was probably the most inaccessible major decorated cave in all of Europe. The crown jewels of prehistoric caves—Lascaux, Altamira, and Chauvet—were operated by the government and periodically opened for researchers and special visitors. Le Tuc, on the other hand, was privately owned: the Bégouëns opened the cave only when they desired, which was at most once a year, and only for prominent archaeologists or friends of the family. Some of the world's most renowned scholars of prehistoric art had waited their entire lives for an invitation to Le Tuc that would never come. Nonetheless, an archaeologist from the University of California, Berkeley, named Meg Conkey, who had worked at Le Tuc, offered to give me the mailing address of Chateau Pujol, so I could at least send a letter to introduce myself to Count Bégouën.

In rudimentary French, I described my preceding years of exploration, told him about REVS, and how my curiosity about hidden artworks had led me to his family's cave.

When I didn't hear back, I wasn't surprised—it had been a long shot after all—and as the weeks passed, I let it go. But then, out of nowhere, I received an email from Count Bégouën. If I could be at Pujol on a particular Sunday in late November at two in the afternoon, he would open the cave for *une visite exceptionnelle*.

As I drove through the gates of Pujol, the chateau came into view: an enormous stone structure with turreted towers set high on a hill with green pastures sloping down on all sides, everything glowing in a van Gogh–ish light. A little apart from the chateau was a smaller rustic stone building that contained the family library, a small archaeological laboratory, as well as a private museum that displayed artifacts discovered on the property.

Count Bégouën emerged from the laboratory and greeted me with a big, buoyant laugh. He was seventy-six, but could have passed for much younger: a tall and lean man, with delicate features, a narrow face, and trimmed dark hair. He carried himself with a straight-backed dignity and had impeccable manners, but was earnest, almost boyish in his warmth—a man who wore grace like a color. I had only my infantile French, and he didn't speak English, but we managed to communicate.

"It's twelve degrees in the cave," he said, as he showed me to a large locker full of caving gear. I stepped into a pair of blueberry-colored coveralls and pulled on rubber boots: "*Tres chic,*" said the count, with a wink.

Welcoming me alongside the count was Andreas Pastoors, a German archaeologist who had been working in the caves on the Bégouën property since he was a teenager.

He was in his mid-fifties, a stern, forcefully intelligent man, with a slightly upturned nose. He was bulldoggishly protective of the caves, and seemed to be baffled by the count's decision to invite me here. As we made our way down the hill in the direction of the cave entrance, he pulled me aside. "This is not an open cave," he said. "Every visit endangers the sculptures. Whatever you write in your book, make it clear that the public are not welcome here."

Le Tuc was discovered in 1912 by the count's father, Louis Bégouën, and his two brothers, Max and Jacques. The boys—teenagers at the time—had been out rambling on the land one day, when they started following the course of the Volp, which led them to the cave entrance. In a small, bathtub-shaped boat, they paddled through the mouth, and spent the afternoon poking around in the first cluster of chambers. For the next two years, every few months, they mounted another expedition, striking each time into slightly deeper chambers, lighting their way with lamps jerry-rigged from bicycle lights. On one trip, the younger Bégouën brother, Max, dismantled a pile of broken stalactites to reveal a new passageway. The brothers shimmied through, followed one passage, then another, deeper and deeper, until they reached the very last chamber, where they discovered the bison. When they emerged, they contacted the two preeminent prehistorians at the time, Abbé Henri Breuil and Emile Cartailhac, who rushed down from Toulouse to see the sculptures. A photograph taken on that day, showing the Bégouën brothers, their father, and the two prehistorians, hangs in the family library.

The Bégouën family cosmos now revolves around the bison. In October of 2012, on the hundredth anniversary of the discovery, Count Bégouën gathered the entire extended family at Pujol for a grand celebration. He taught the young Bégouëns how to throw spears and make stone tools and build friction fires like the Magdalenians, and then he recounted the tale of their ancestors' discovery of the clay sculptures. In the evening, the family came together for a feast—the main course was bison, the Bégouën family totem.

"Everything is now as it was in 1912," said Count Bégouën, standing on a mossy boulder at the mouth of Le Tuc, with the Volp flowing around him. He wore blue coveralls, a white helmet, a miner's lamp, and, over one shoulder, a leather satchel delicately tooled with his father's name: he looked like the hero of an old silent film.

"Even the boat is the same," he said, pointing to the little bathtub-shaped vessel in the water, a near-perfect replica of the one used by his father and uncles.

WE WERE SIX IN ALL, including three members of Andreas's archaeological team: Hubert, an older archaeozoologist from the University of Cologne, who studied bones of prehistoric animals; Julia, a student of Hubert's; and Yvonne, who worked with Andreas at the Neanderthal Museum. They'd all been at Pujol for the past two weeks, studying artifacts excavated from another cave on the Volp called Enlène. For all three, it would be their first visit to Le Tuc. We boarded the boat two by two, paddling upstream, wobbling into darkness, before making landfall on a small beach of gravel. From there, we adjusted our lamps and turned in to the cave.

Le Tuc twists upward into the limestone hill, tracing the ancient route of the Volp. The American poet Clayton Eshleman, in his poem "Notes on a Visit to Le Tuc d'Audoubert," which he wrote after a visit with Count Bégouën in 1982, described the cave as "the skeleton of flood." Over the course of the afternoon, as we trekked to the bison sculptures, we would ascend through three distinct levels. From the riverbed, we climbed up into a chamber known to the Bégouëns as *La Salle Nuptiale*, "the Wedding Room": a broad, vaulted chamber, where large stalactites hung from the ceiling like the pipes of a church organ.

Andreas addressed the group in English, which would be the lingua franca of our trip. Le Tuc is a pristine cave, he explained. Unlike most archaeological sites, where the floor is ripped apart during excavation and artifacts are removed to laboratories or museum vitrines, Le Tuc is essentially untouched, with almost every trace of the prehistoric visitors left in situ. "Always try to put your feet where I put mine," said Andreas. "And never, ever touch the walls."

We began our hike, ducking and angling our bodies to avoid stalactites, stepping gingerly and quietly on the soft clay floor, the way you would if you were sneaking up on someone. Which, in a way, was just what we were doing. We were on the trail of the Magdalenians, who last entered this cave 14,000 years ago. The Magdalenian culture, which dates from 17,000 to 12,000 years ago, is the darling of European prehistory. The two preceding cultures—the Solutreans (17,000 to 22,000 years ago) and the Gravettians (22,000 to 32,000 years ago)—certainly had their own moments of brilliance: they sculpted elegant stone tools, carved beautiful portable statues, such as the ample-bottomed *Venus of Willendorf,* and painted the breathtaking images in the celebrated cave of Chauvet. But the Magdalenians were the virtuosos—the Renaissance Florentines of the Paleolithic Age. They painted the reindeer and bison in Lascaux and Altamira that early archaeologists swore were a hoax because they were simply too finely executed to be ancient. Along the backs of rock shelters, they chiseled sculptural friezes of galloping horses, and crafted musical instruments from bone, which employed exactly the same pentatonic scale that we use in our instruments today. They sewed their clothing with bone needles and adorned themselves with fine necklaces strung with seashells. Even their most utili-

tarian tools were delicately embellished, such as one antler-carved spear thrower, which was etched with the form of a bison.

Not long into our trek, as we stooped beneath a curtain of stalactites, Andreas motioned for us to pause. Angling his light down at the soft earth, he revealed a fossilized footprint. Visible in perfect relief were each individual toe, the gentle curve of the arch, the cup of the heel. It was a poignant reminder that we and the Magdalenians, however separated by time, were physiologically the same: we had the same bodies, the same brains, the same nervous systems, the same essential ways of being in the world.

As we clambered deeper and deeper into Le Tuc, it became clear how much more perilous the same journey would have been for the Magdalenians. Where we ascended a steep cliff-face on an iron ladder installed by the Bégouëns, our predecessors would have made the same climb sans iron ladder; they would've done it barefoot, and, according to a marking in the cave identified as the print of a bare knee, without pants.

Farther along, Andreas passed his light over an enormous print in the clay, of a cave bear claw. It was an arresting encounter for us, but ultimately abstract, as cave bears had gone extinct thousands of years before. The Magdalenians, however, lived contemporaneously with cave bears: seeing these claw marks etched in the mud would have left them cold.

Soon we came to the place where, in 1912, Max Bégouën had unclogged a natural breakdown of stalactites, opening a passageway deeper into the cave. At the *chatière*, as the family called it, the cave sharply tapered, as though cinched by a belt, becoming a long, narrow, tight passageway, which could be traversed only by way of an extended horizontal belly-crawl. We paused a moment as the four of us first-timers eyed the passage. In the past, I'd been told, certain unslender visitors had come to the *chatière* and been forced to turn back. "Just make sure you breathe easily," Andreas told us. "And if you do feel uncomfortable, absolutely do not try to get up. Stay down."

The count went first: a man just south of eighty, wriggling on his elbows through a narrow, rocky aperture. (*"C'est un bon exercice, cette cave,"* he said, modestly.) After watching the treaded soles of Julia's boots disappear around the corner, I started my crawl. The passage was tight, and scaled with small stalactites. I snaked my hips and dragged myself on my elbows. One of the earliest researchers to visit the cave, the German anthropologist Dr. Robert Kuhn, described his trip through the *chatière* as "creeping through a coffin." Eshleman recorded "the dread of withering in place." A panel of engravings along the *chatière* showed a string of phantasmagoric figures—the Bégouëns call them *les Monstres*—who seemed to be guarding the passage.

Out of the squeeze, we emerged into a room of calcite growths so pure and delicate it looked as though a winter ice storm had blown through, encrusting everything in a pearly freeze. When Louis Bégouën and his brothers entered this room, where no human had stepped foot in fourteen thousand years, the mere sound of their voices had caused the glasslike stalactites to pop and burst around them, the shards tinkling to the cave floor.

We traversed another grove of stalactites, weaved around a large floor-to-ceiling pillar, stooped up a shallow incline, then down again. We'd now been hiking for two and a half hours: everyone had gone quiet, all of us floating in a detached trance. And then, all at once, Andreas dropped to his knees on the path and motioned for the rest of us to kneel beside him. His body went still.

"And now I will ask you"—he said this softly and slowly, as though delivering a line from a dimly lit stage—"to turn off your lamps." As I grasped what was about to happen, my heart quickened. Total darkness fell over the chamber, and we all went silent. Then Count Bégouën's lamp flickered on, and he angled the beam into the dark behind us. Like a line of puppets on a single string, we all turned to follow the beam, and in that moment, in perfect unison, we all froze.

It was a small room with a domed ceiling, the floor flat and bare: at its center, perhaps ten feet from where we crouched, was a large stone. Leaning against the stone, tilted slightly away from us, and glowing in the soft lamplight, were the clay bison. I could hear my companions all exhale in unison. I felt my whole body go tense, tendon by tendon, my muscles tightening and bunching around my shoulders. And then, all at once, everything came unbound: a warm tide welled up inside of me, rising from my core, through my torso and my shoulders, then up into my head, until my breath went ragged. All at once, as I peered at the bison, I began to sob, tears stealing down my cheeks.

The front bison was a cow, the one behind a bull. The granularity of detail was exquisite: the swoop of the bisons' horns, the ridge of the chin, the cascade of the beard, the gentle slope of the hump, the crest of the belly, the bulky hunch of the shoulders. You could almost see muscles contracting, organs palpitating under the skin. The clay gleamed, as though it were still damp. The sculptures looked like something out of a children's story, where dolls and puppets surreptitiously come to life when the lights go out.

Count Bégouën slowly led us crawling in a soft circle around the sculptures. He quietly began to point out details in the sculptures, speaking tenderly, as though introducing us to a member of his family. At first, he spoke slowly, using simple phrases in French, which I was able to follow without much trouble, but as he pointed out increasingly minute details in the clay, he began to speak more quickly, until I eventually lost the thread, and found myself receding from the conversation. As I let my eyes soften, the details of the bison began to fall away, and I pulled back as the count's voice sounded farther and farther away. Soon I felt I was alone in the chamber, the only audible sound the blood in my ears.

The Bégouëns, I'd been told, kept a leather-bound *livre d'or*, or guest book, in which a century's worth of visitors to the cave had recorded their gratitude to the family and inscribed accounts of their experience of glimpsing the bison. Nearly everyone had been cast into a similar swoon. Louis Bégouën described being "nailed in place, unable to speak." Conkey, too, described a sensation of paralysis. When Dr. Kuhn arrived at the back of the cave, he felt a "redemption" wash over him. The poet Clayton Eshleman wrote, "In Le Tuc d'Audoubert I heard something in me whisper me / to believe in God." Everywhere in the *livre d'or* was the language and cadence of spiritual awe. Every visitor to this chamber had felt *mysterium tremendum et fascinans,* or "fearful and fascinating mystery," which is how the philosopher Rudolph Otto described the basic elements of the holy.

Which is strange. After all, we know almost nothing about the bison. Fourteen thousand years later, the Magdalenians are shadows in the dark: we know about their lives only what we've pieced together through bones and the scattered ash of ancient campfires. We can make only the haziest speculations about their myths and gods, the shapes and contours of their cosmos. Any sacred objects, wrote the sociologist Robert Bellah, must be regarded "in the context of a community for which they are sacred: it is in ritual practices of a living community that they become sacred." Any cultural context of the bison—the practices that would have infused them with significance for the Magdalenians—is irretrievably lost. And yet, one hundred and forty centuries later, when people of today's world visit this chamber—people who drive station wagons and buy frozen food at grocery stores—they fall to their knees before the sculptures. Even now, we all lay prostrate before the bison in the dark, peering up at them in a posture of worship, our eyes radiant and damp. In this chamber, time collapses, the space between ourselves and our ancestors narrows to a hair's breadth.

"THE WORDS SECRET AND SACRED are siblings," wrote the poet Mary Ruefle. At the back of Le Tuc d'Audoubert, what we feel is the interwoven power of sacredness and hiddenness—secrecy and holiness, concealment and divinity—which runs through all spiritual practice. It is central to Hindu worship, in which a devotee at a temple approaches a statue of a deity enclosed in a dark chamber. And to the initiation rites of the Urapmin tribe of Papua New Guinea, which are conducted in a mysterious and inaccessible enclosure kept forever in pitch-darkness. And even in ancient Egypt, where the holiest compartment of a temple is its darkest chamber, a secret sanctuary nestled behind stone thresholds.

"The Lord," says King Solomon in the Old Testament, "has set the sun in the heavens, but has said that he would dwell in thick darkness." Indeed, all Abrahamic religion is rooted in the idea of sacred hiddenness, going back to the Tabernacle, which was the original template for all sacred architecture. A portable tentlike structure, the Tabernacle served as God's sanctuary as the Israelites wandered the desert after their exodus from Egypt. It comprised an open courtyard, at the center of which was a rectangular tent, with a door leading into a compartment that only the priests could enter. At the back of this compartment, hidden behind a veil, was an enclosed chamber, called the Holy of Holies, or the Sanctum Sanctorum, which was perpetually kept in perfect darkness. Here lay the most sacred relics, including the Ark of the Covenant, the ultimate representation of God. Only the High Priest was allowed to enter the Holy of Holies, and only once a year, on the Day of Atonement, when he sprinkled blood on the Ark to atone for humanity's sins.

When the Hebrews finally arrived in Israel, they built the first temple on the Temple Mount in Jerusalem, following the strict blueprint of the Tabernacle: the Holy of Holies was located in a cave within the depths of the stony hill. No archaeologist has ever conducted a full exploration of the chamber—it is too sacred, too politically divisive—but it is said that if you tap the floor in the right place, you can hear an echo reverberating from below. In any case, it's not hard to imagine the Tabernacle and the Holy of Holies as an architectural reproduction of a cave, a dark zone chamber made portable, which enabled the wandering Israelites to conduct versions of ancestral ceremonies that were long ago performed in subterranean darkness.

IF WE CAN CAST any light, even just a glowworm's worth of light, on what happened in the depths of Le Tuc so long ago, it may just come from a chamber adjacent to the room of the bison, where Andreas now led us. Crouching on the threshold, he cast his light over a large pit in the center of the chamber, from which the Magdalenians dug up the clay for the sculptures. Scattered throughout the chamber, Andreas said, as he darted his light over the floor, was an array of 183 human footprints, nearly all of which, bafflingly, were heelprints. In 2013, in an effort to solve the mystery of the heelprints, the Bégouën family arranged a visit to the cave with three men from the San tribes of the Kalahari Desert. The San are genetically the oldest culture in the world and one of the last extant groups to follow a traditional hunter-gatherer lifestyle. These men were expert trackers; by looking at a clear footprint, they could determine the person's sex, whether they were injured or sick or carrying something, whether they were walking quickly or slowly, even whether the person was frightened or relaxed. At the back of the cave, the trackers spent an hour crouching over the tangle of heelprints, animatedly clicking back and forth in their native Ju/'Hoansi.

Fourteen thousand years ago, they concluded, two people had been present in the chamber: a boy of about fourteen years old, and a man of approximately thirty-eight. They had walked back and forth, excavating large chunks of clay from the pit, then hefting them into the room of the bison, the weight of the burden causing their feet to sink into the mud. But the heelprints, the trackers asserted, were not the result of maneuvering in a low-ceilinged room: the Magdalenians had walked on their heels intentionally. It seemed plausible, they said, that the prints were the remnants of a ritual—a dance, perhaps—though it was difficult to say for sure. In any case, the trackers recognized the practice of heel-walking. In the Kalahari, they explained, where everyone in a given group knows one another's tracks, leaving your footprint in a place is like signing your name. (It is impossible, for example, to have an affair within a group, because any footprint immediately betrayed a late-night rendezvous.) The only way to conceal your identity, to be truly anonymous, is to walk on your heels. As we crouched over the fossilized heelprints, I imagined the Magdalenian sculptors working by dim torchlight, preparing for a ceremony we will never fully understand, but which was so incandescently sacred that they had to obscure their identity, to hide themselves even in this hidden place.

As we retreated to the entrance of the cave, worming back through the chatière, then scrambling back down the cliffs and into the Wedding Room, I recalled a story I'd heard about Pablo Picasso visiting the cave of Lascaux, which is only a few hours' drive from Le Tuc, and was also painted by the Magdalenians. This was in 1940, just months after the cave had been discovered, at a time when it was still raw, not yet outfitted for tourists, with only a ragtag group of locals leading visitors underground by torchlight. Picasso climbed down into the damp cave, watching as his guide cast soft, guttering light over the ceiling, revealing bulls and reindeer and horses galloping across the stone. In that moment, as Picasso was overwhelmed by a collapse of time, where the tide of antiquity washed over the present, he whispered, "We have invented nothing new."

AT THE END OF the evening, we met in the Bégouën family library, gathering around a heavy mahogany desk, which displayed a bronze model of the bison sculptures and a photograph of Count Bégouën's grandfather. It had long grown dark, everyone was tired, muddy around the ears, but still glowing from our encounter in the cave. We all signed the *livre d'or,* and Count Bégouën opened a bottle of muscadet, which we drank from plastic cups, toasting to the afternoon, and to the bison.

The sculptures, the count said, were "a miracle of preservation." Had they been placed just a few feet to the left, or a few feet to the right, he said, the bison might have been destroyed by water dripping from the ceiling, and lost to history. While any taphonomist—a scientist who studies the nature of decay—would confirm that the bison's preservation was remarkable, I wonder if *miracle* is the right word. I suspect the Magdalenians would have understood patterns of decay and preservation. For a people who lived in nature's perpetual evanescence, whose daily lives were entirely tethered to the vicissitudes of shifting weather, movements of prey animals, and the sprouting of seasonal vegetation, they would have recognized the unique power of a space that was truly inert. They would have known that to place something in the sealed subterranean depths was to make it last forever.

As I made my way back through the front gates of Pujol, I thought of the bison, secreted away in their small domed sanctuary, and I marveled at the likelihood that, in another fourteen thousand years, regardless of how the world had changed on the surface, they'd remain in the same pristine condition, as though encased in amber.

BACK IN NEW YORK, roughly ten years after I'd begun looking for him, I found REVS. One afternoon, I was talking to my friend Radi in the restaurant he owned in Brooklyn. He was telling stories about his family, how his father had come over from Palestine, and sold costume jewelry up and down Broadway, until he made enough money to buy a deli, which he then expanded into a grocery store, and how he eventually bought a house in Bay Ridge, where Radi lived as a boy.

Years ago, I told Radi, I'd spent time in his old neighborhood, where I'd gone to search for a graffiti artist, a ghostlike man called REVS, who'd written his life story in the hidden parts of New York.

"I know REVS," said Radi, grinning. "I grew up with REVS."

ON A COLD, GUSTY NIGHT in February, in a scene that I'd imagined so often it felt like a memory even as it was happening, I found myself at a pizza dinner in Brooklyn, sitting across from REVS. Around the table was a mingling of painters, musicians, and filmmakers. REVS was in his early fifties, with boyishly red cheeks, and slate blue eyes beneath a wool beanie. While everyone else joked and laughed, he sat back quietly in his chair. He was guarded, wary of everyone, and especially of me, the only person he didn't know.

I wanted to tell REVS that his diary was part of why I'd fallen in love with New York, that I had transcribed almost every page, that I used to quote his words to friends, and probably knew the text better than anyone in the world outside of him. In preparation for the dinner, I'd filled a notebook full of questions about specific quotes in the diary, imagining that we might do a kind of close reading of his book. But the moment I brought up the diary— I nonchalantly asked how he'd gotten the idea—it was clear he did not want to talk about it. He shifted in his chair and crossed his arms.

"I was on a mission," he said. With that, he turned back to his pizza, signaling the end of the conversation.

I tried to prod him, but again he deflected. "I just got on a mission. That's all I really want to say about it," he said.

Dinner continued, conversation drifted through old stories of New York, about old graffiti gangs battling one another for territory. REVS jumped in with a word every once in a while—he still spoke, I noticed, in the slang of a teenage graffiti writer from the eighties—but otherwise he kept to himself.

Finally, during a lull in the conversation, I caught REVS's eye. Feeling emboldened, I told him that I'd gone down in the tunnels, and run the tracks, and that each time I found one of his pages in the dark, I'd felt a private thrill.

He squinted at me—still gruff, but now just a hint less dismissive.

I wanted to ask REVS about one of the pages I'd seen—number 80—where he described the beginnings of his life in graffiti. "I was struck by the idea of leaving your mark on something," he wrote, "and if I could, too, I'd last forever." I wondered if he imagined a future version of New York, when the city is crumbling and overgrown, and a group of wary explorers climbs underground, lighting their way through moldering tunnels, until they discover one of his pages, still preserved in the dark.

"Was the diary about making something that would last?" I asked.

He shrugged, without saying anything.

After a moment, though, REVS turned to me and said, "You know, some of the pages are sealed."

I looked up.

"Back in the day, I'd paint on the back walls of emergency exits," he said. "A few of those alcoves are sealed over now with bricks."

"Who sealed them?" I asked. "The MTA?"

I imagined a solitary figure down in the tunnels late at night, a man dressed in a stolen MTA hardhat and fluorescent vest, holding a spade and crouching over a bucket of concrete, laying brick by brick, obscuring his own paintings in the dark.

REVS held my eye for just a moment, then looked away.

CHAPTER 8

THE
DARK ZONE

To know the dark, go dark.
Go without sight,
and find that the dark, too,
blooms and sings.

—WENDELL BERRY,
"TO KNOW THE DARK"

On July 16, 1962, high in the French-Italian Alps, a twenty-three-year-old French geologist named Michel Siffre strapped on a helmet, nodded solemnly to a small ring of friends and well-wishers, then climbed a wire ladder down through the mouth of Scarasson Cavern. He touched down in complete darkness, some four hundred feet underground, and cast his light beam over the interior of the cave, where the walls gleamed with dense blue ice. Waiting for him in the central chamber was a red nylon tent, several pieces of foldable furniture, a large store of canned food and water, and a one-way field telephone with a wire running to the surface. Siffre signaled to his surface team with a tug on the ladder, then watched as it slowly receded from view, until he was entirely alone in darkness and silence. Siffre would live in this subterranean hollow, in total isolation, without coming to the surface, for the next two months.

It was an experiment in chronobiology: the study of life's innate biological rhythms. The idea was that in the vacuum of the cave—where Siffre was in absolute darkness, cut off from sunrise and sunset, with no access to a calendar or watch—his body would revert to a natural sleep-wake cycle, a primordial body clock. He would discover, he said, "the original rhythm of man."

During his stay in the cave, Siffre would discern the passage of time entirely on instinct. He would track his schedule in a log: each time he felt drowsy and prepared to sleep, each time he awoke, and each time he ate a meal, he would record what time he *felt* it was; in turn, he would report his schedule by telephone to his support team on the surface, who would keep track of the objective time. Outside of these exchanges, any communication between Siffre and the support team, which comprised a few of Siffre's classmates from the Sorbonne, was forbidden, lest they reveal any hints about the time of day on the surface. At the conclusion of the experiment, Siffre would compare his subjective underground chart, where his unit of time was "awakenings," with the objective time chart on the surface, and see where they diverged.

As the ladder rattled out of view, Siffre became a full-time resident of the dark zone. He had a few weak flashlights and a carbide lamp, but in order to conserve batteries and gas, these were kept off much of the time. Siffre spent his days—or "days"—listening to Beethoven sonatas on a record player, reading by flashlight (Tacitus, Cicero, and a few adventure survival books; he'd meant to bring Plato's *Republic*, with its "Allegory of the Cave," but had forgotten it at home). He dreamed about his girlfriend in Paris, or played a game in which he tried to throw lumps of sugar into a boiling pot of water in pitch-darkness. At one point, he befriended a spider, which he kept in a small box. ("She and I are all alone in here," he wrote in his journal.)

As he followed his routine—sleep, wake, sleep—he found himself grappling with the caprices of the subterranean environment, a world of "unrelenting sameness." As the days passed, he slipped into a hibernatory torpor, where his metabolism slowed, his visual and auditory faculties dulled, and his mind gradually came unmoored. He was haunted by the "terrifying sensation of infinite space" and wondered what had driven him to this undertaking: "Surely I hadn't embarked on this expedition of my own free will," he wrote. "Surely some outside and superior force had impelled me!" He began hallucinating, seeing flashing spots in front of his eyes. At one point, he found himself shrieking into the darkness. "I now understand," he later wrote, "why in their myths people have always situated Hell underground."

On September 14, Siffre's sixty-third day in the cave, the support crew dropped the wire ladder through the cave mouth, announcing the conclusion of the experiment. Siffre was baffled: according to his own chart of "awakenings," it was August 20. His perception of time had fallen a full twenty-five days behind. Curiously, according to the log kept on the surface, his body had not lost track of time: Siffre's average sleep-wake cycle hovered around twenty-four hours.

Siffre dedicated his entire life to studying his natural biorhythms in the depths of caves. A few years after the mission at Scarasson, the "Jacques Cousteau of the Underground," as he became known, cloistered himself deep inside a cave near Cannes. Then, on a NASA-sponsored expedition in 1972, he lived in isolation for six months in Midnight Cave in Texas. At age sixty, he spent two months alone in a cave in France called Clamouse. In almost every case, there was a moment when his mind turned on him, in which he experienced phenomena disconnected from reality. As I studied Siffre's work, reading his reports of each experiment, I sensed that there was something more to these expeditions than sleep cycles, that his extended retreats into darkness touched something stranger and more elemental.

In *Beyond Time,* the book Siffre wrote about his maiden experiment at Scarasson, I came upon a photograph of the young scientist emerging on the final day of the experiment. Following two months in "eternal subterranean night," he was too weak to make the climb himself, and so was hoisted to the surface in a parachute harness. Siffre dangled, as slack as a marionette, flickering in and out of consciousness. The blacked-out goggles he wore to shield his eyes from the sunlight suggested a kind of cosmic travel. Unrecognizable from the man of two months before, he

was pale, hollow-cheeked, skeletal, as though he'd died and was being raised back into the world of the living.

The image of Siffre reemerging from Scarasson reminded me of a story about Pythagoras, the ancient Greek

philosopher, who was known for his own extended retreats
into caves.

Pythagoras is today known as a mathematician, but
during his life in the sixth century B.C. he was celebrated as
a semi-divine man, a sage, who could hear "the music of the
stars," as one contemporary wrote. While none of his writings survive, his followers relay that Pythagoras healed the
sick with incantations, predicted earthquakes, suppressed

thunderstorms, journeyed into the past, and had the capacity to "bilocate," meaning he could be in two places at once.
Even allowing for a generous margin of exaggeration, no
one doubted that Pythagoras possessed, in one form or another, a measure of extra-human potency. (Even sensible
Aristotle acknowledged, "Among rational creatures there
are gods and men and creatures like Pythagoras.") Everyone agreed, too, that part of the source of Pythagoras's wisdom was his practice of enclosing himself in dark caves for

extended periods of time. He possessed his own cave on Samos, which he called his "House of Philosophy," where he would retreat into darkness, in order to meditate on the intricacies of the cosmos. Once, Pythagoras wrapped himself in a black lamb's wool and descended into a cave on Crete and did not reemerge for twenty-seven days. When the philosopher finally lurched out of the dark, pale and haggard, he announced to his disciples that he had experienced death, that he had journeyed to Hades and returned and now possessed sacred knowledge beyond any mortal rhythm.

I wondered about these parallel cave retreats: Siffre, who journeyed into the dark to test his biological limits; and Pythagoras, who descended in search of mystical wisdom. The two seemed to be speaking to each other, sharing a secret across two thousand years. It was my curiosity about this echo that led me to conduct an unscientific and perhaps reckless experiment. I would perform my own dark-zone retreat. I would camp out at the bottom of a cave and spend twenty-four hours alone in uninterrupted darkness.

To help with the logistics of the experiment, I consulted a caver friend in New York named Chris Nicola. One of the more experienced cavers in the United States, Chris had explored caves in dozens of countries around the world. He'd also spent an inordinate amount of time thinking about prolonged stays in the dark zone. In 1993, while exploring a gypsum cave in western Ukraine called Priest's Grotto, Chris discovered the remnants of an old camp, situated seventy feet underground. He found wooden bed frames, broken ceramics, porcelain buttons, a millstone to grind flour, and a dozen leather shoes. Chris spent years unraveling the mystery of the camp at Priest's Grotto. During World War II, he learned, thirty-eight Ukrainian Jews, including elderly women and young children, had lived in the cave for a year and a half, hiding from Nazis. He tracked down all the survivors from the cave, interviewing them about their experiences in subterranean darkness, then recording their stories in a book called *The Secret of Priest's Grotto*, and a documentary titled *No Place on Earth*.

When I told Chris I wanted to explore the effects of the dark zone on the human mind, he knew precisely what I was talking about. He even knew where I should go to conduct my experiment. He had a longtime caver friend, a man named Craig Hall, who lived in Pocahontas County, West Virginia. Craig owned a big plot of land down there, Chris said, that was riddled with caves.

"Craig knows cave darkness pretty well himself," he said. "Go find him. He'll set you up."

ON MY DRIVE THROUGH West Virginia—sidewinding hill roads, old cockeyed cabins, bait shops, country churches— I swore I could smell the cool, fresh musk of cave air wafting out of the trees. West Virginia is cave country: nearly the entire state sits on karst, the limestone topography that is so easily hollowed out by water to form caves. According to the National Speleological Society, West Virginia is pocked with about 4,700 caves, or 5.1 caves per square mile, the highest density in the country. When I stopped for a sandwich at a little grocery store near Craig's home in the town of Hillsboro, the old couple behind the counter asked what had brought me down from the North. I told them I was going to visit a man who owned some caves.

"You own land of any size here," the man said, "you own caves."

Craig Hall met me in his driveway, which was obscured behind a wall of trees up in the hills. He was a tall, slender man in his mid-sixties, with long, ropy limbs that had served him well over some forty years' worth of caving. He wore his hair pulled back in a messy gray ponytail, the way American frontiersmen appear in old paintings. He and his wife, Tiki—a short, sharp-witted woman, also a caver— lived on a two-hundred-acre tract of wild, unkempt woodland, in a two-story home dwarfed by towering oaks. He and Tiki met on a hippie farm in North Carolina in the early 1970s, then got into their VW bus, drove into rural West Virginia, liked what they found, and stayed. It had been forty years since they arrived: long enough that it was home, but not so long that they were numb to the gothic notes of Appalachia. There were old families down here, Craig told me, that were so isolated and had been here so long, they still spoke with the Irish accents of their ancestors. The next county over, he said, there was a family of reputed murderers, all of whom had double rows of teeth. Friends of his had seen ghosts in these hills, young soldiers in Confederate uniforms, holding muskets, marching through the forest.

"There's been some rain, so most caves here are flooded," Craig said. "But I have one that might do you well."

As we hiked up to the entrance of Martens Cave, we felt a sigh of cool air coming from the darkness. The cave, Craig explained, was about a quarter mile long, and had a stream flowing through the middle. As a walk-in, walk-out cave, he explained, it was easily accessible to me, but also accessible to animals. Standing at the cave mouth, Craig checked off a roster of the creatures I might encounter. Raccoons ("they're always around"), bears ("not so much this time of the year, but maybe"), pack rats ("if you see a little bundle of leaves, that's them"), bobcats ("probably"), panthers ("yup"). He paused, perhaps noticing that I'd gone slightly wan. All things considered, he said, I shouldn't worry. "Remember, human doesn't taste good. Don't bother them, they won't bother you."

It was a little after 6 P.M. If Craig didn't see me at his

house by about this time tomorrow, we agreed that he'd come looking for me. With that, Craig headed back to his truck, and I headed into the dark.

COMPARED WITH SIFFRE'S IN Scarasson, my camp in Martens Cave was luxurious. I settled down a few hundred feet from the entrance, on a stretch of soft, dry earth, where the ceiling was high enough that I could stand. The temperature was about 55 degrees Fahrenheit, and the stream, which passed about twenty feet from my camp, emitted a soft chuckle. I laid out my sleeping bag almost flush against the cave wall, where I figured a panther would be unable to ambush me from behind. When I cast my lamp upward, droplets of condensation on the rock ceiling shimmered celestially.

I downed a sandwich, followed by a pull from a bottle of moonshine, a good-luck gift from a West Virginian caver friend. Then I peed in the stream, sat down on my sleeping bag, and looked at my watch—6:46 P.M. I braced myself, took a deep breath, reached for my headlamp, and killed the light.

AT FIRST, THE DARKNESS wasn't such a shock. Not so different, I thought, from waking up late at night in an unfamiliar room and waiting for your eyes to adjust. I leaned back against a small boulder, pulled my sleeping bag over my lap. I let out a small moonshine burp. I felt a glassy calm. I sat cross-legged, straightened my back, and stared into the

dark; for those first few moments, I focused on my breathing, felt my thoughts fall away, and imagined I could sit for
days. What changed everything is that I blinked. I blinked
and saw that I could detect zero evidence that I'd blinked. I
could feel the *act* of the blink—muscles twitch, eyelids slide
down, eyelashes brush together, eyelids slide up—but de-
tected no result. It felt like my body and brain were not
communicating, like a power line had gone down in a
storm.

Our aversion to darkness is rooted in our eyes. We are diurnal—day-active—creatures, meaning our ancestors, down to the finest physiological points, were adapted to forage, navigate, and seek shelter while the sun was up. Sure enough, by daylight, our eyes are magnificent. We possess an abundance of the photoreceptor cells called "cone cells" that enable us to home in on sharp details: our ancestors could pick out game animals on the horizon, or glimpse a piece of fruit in a tree and know from the precise shade of color whether or not it was ripe. But without sunlight, our eyes are all but useless: for our glut of cone cells, we lack the other type of photoreceptor—"rod cells"—which enable vision in low light. As the sun set each night, our ancestors became vulnerable, transforming from predator to prey as they entered a world dominated by nocturnal hunters, all endowed with powerful night-vision: lions, hyenas, saber-toothed tigers, venomous snakes. For our ancestors, the height of terror would have been wandering over the savannah in the dark, listening for a predator's paws drumming against the ground.

In the modern West, we no longer fret over night ambush by saber-toothed tigers, but we still squirm in the dark. "After thousands of years," wrote Annie Dillard, "we're still strangers to darkness, fearful aliens in an enemy camp, with arms crossed over our chests." I have many times been disquieted by darkness. In childhood games of sardines, hiding in the corner of my father's closet, my heart thumping. In the bush in Australia, getting up to pee without a flashlight, losing sight of the tent, stumbling through the dark, thinking of packs of dingoes. After Hurricane

Sandy in New York, walking through lower Manhattan, down block after block of the city's blacked-out grid, hair bristling on the back of my neck. But these were partial darknesses, always with a dot of light through a keyhole, or a star-gleam from the sky. Here the eye will always adjust, the iris will always open to collect photons. Not underground. In cave darkness not a single photon penetrates. Here lies a heavy, ancient dark, Book-of-Genesis dark.

My thoughts earthwormed down inside my body, chewing through my inner architecture. It was the feeling of being peeled open, turned inside out. I felt the rhythmic clenchings of my heart, my lungs ballooning inside my ribs, my epiglottis flapping open and shut. In the absence of sight, my other senses bloomed. The sound of the stream, which I'd barely noticed when I entered the cave, now filled the whole chamber, unfurling in effusive patterns. Smells—mud, damp limestone—thickened to the point of feeling material. I could taste the cave. When a drop of water fell from the ceiling and burst on my forehead, I almost jumped out of my sleeping bag.

OUR FIRST STUDIES ON sensory deprivation grew out of a clandestine Cold War military experiment on mind control. In the early 1950s, footage emerged from Korea of American prisoners of war denouncing capitalism and extolling the virtues of communism. The CIA, convinced that the soldiers had been brainwashed, promptly launched a research initiative—Project Bluebird—on mind-control techniques. Part of the research team was a psychologist named Donald Hebb, who offered to conduct an experiment on what he called "sensory isolation."

Hebb was not so interested in actual brainwashing, but had long been curious about the response of the brain to the absence of stimuli. He wondered about reports, for example, of Royal Air Force pilots who, after many hours of flying in isolation and staring at an unchanging skyline, would suddenly, for seemingly no reason, lose control of the plane and crash. And of mariners who, after stretches looking out over a static marine horizon, saw mirages. And of the Inuit who warned against fishing solo, because in the absence of human contact, without visual cues in a white-out Arctic landscape, they would become disoriented and paddle out to sea, never to return. By examining the neurological response to isolation, Hebb wondered if he might be able to answer questions about the structure of the brain.

For Project X-38, Hebb constructed a grid of four-by-six-by-eight-foot cells, each air-conditioned and sound-proofed, then recruited volunteers, whom he paid twenty dollars a day to lie in the cells, where they were subjected to "perceptual isolation." Over their eyes, the subjects wore frosted plastic goggles that prevented "pattern vision." To reduce tactile stimulation, they wore cotton gloves and elbow-to-fingertip cardboard cuffs. Over their ears, a U-shaped foam pillow. The cells were outfitted with observation windows, as well as an intercom so that the research team could communicate with the subjects. Hebb instructed his volunteers to stay in the cells for as long as they could.

Initially, Hebb had regarded Project X-38 lightheartedly, joking that the worst part of isolation for the subjects would be the meals prepared by his post-docs. When the results came in, though, he was stunned: the subjects' disorientation was far more extreme than he'd imagined. One volunteer, upon completing the study, drove out of the laboratory parking lot and crashed his car. On several occasions, when subjects took a break to relieve themselves, they got lost in the bathroom, and had to call a researcher to help them find the way out.

Most startling were the hallucinations. After just a few hours in isolation, nearly all the subjects saw and felt things that weren't there. First they would see pulsing dots and simple geometric patterns; these grew into complex isolated images floating about the room, which then evolved into elaborate, integrated scenes playing out before the subjects' eyes—"dreaming when awake," as one participant described it. One participant reported seeing a parade of squirrels marching "purposefully" across a snowy field, wearing snowshoes and backpacks, while another saw a bathtub being steered by an old man in a metal helmet. In a particularly extreme case, a subject encountered a second version of himself in the room: he and his apparition began to blend together, until he was unable to discern which was which. "It is one thing," wrote Hebb, "to hear that the Chinese are brainwashing their prisoners on the other side of the world; it is another to find, in your own laboratory, that merely taking away the usual sights, sounds, and bodily contacts from a healthy university student for a few days can shake him, right down to the base; can disturb his personal identity."

Today, the neurological mechanism behind these reactions is more or less understood. At any given moment, our brain is receiving a torrent of sensory information—visual, auditory, tactile, and so on. We are so accustomed to this stream of input that when it gets cut off, our brain essentially produces its own stimuli. It identifies its own patterns, combining any scant blip in the visual cortex with images stored in memory to devise scenes that may be intensely vivid, however disconnected from present reality. During one particularly illuminating experiment in 2007, researchers at the Max Planck Institute for Brain Research in Frankfurt collaborated with a German artist named Marietta Schwarz, who had volunteered to live with a blindfold on for twenty-two days. *Blindversuch* (Blind Study), as Schwarz called her project, was part of a larger art project called Knowledge of Space, which incorporated interviews with blind people on perception, image, space, and art. Schwarz sat blindfolded in the laboratory, recording into a Dictaphone a granular, real-time diary of everything that was happening in her brain. She reported an array of hallucinations, including intricate abstract patterns, such as bright amoebas, yellow clouds, and animal prints. The researchers, meanwhile, used an fMRI scanner—functional magnetic resonance imaging, which tracks changes in blood flow in the brain—to follow the neurological operations behind her hallucinations. Despite the total absence of visual input, Schwarz's visual cortex lit up like a lantern, exactly as it would if she weren't blindfolded.

That is, in the world of her brain, the hallucinations were as true and real as anything she could touch or taste or smell.

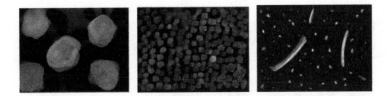

I'D BEEN DOWN IN the dark maybe two hours when they emerged. Just above my head, small glowing orbs of light, ringed with faint areolae, moving in a soft, pulsing dance. They appeared slowly and gently, as though someone in the distance were quietly beginning to sing. I lay in the dark, trying not to move, or even breathe, as though they were wild creatures, and a sudden movement might startle them and cause them to scatter. The lights sent my mind slowly gyrating down into memory, turning up, then dipping down again. I'm a boy on a roof in Providence on a predawn morning, watching a meteor shower flicker gracefully in the sky. I'm eighteen on a boat in a lagoon in Costa Rica, seeing dots of luminescent plankton floating on the sea. I'm alone on the plains of central India, tracking lightning bugs in a rippling cloud. In the rational surfaces of my mind, I knew these orbs were phantoms, the product of an aberration in my nervous system, and yet they were so vividly *present*: growing and shrinking, clustering and unclustering. As the lights brightened, I felt an uncanny weightlessness, as though I were falling gently through space. The lights grew brighter still and I was sud-

denly aware that my body had gone rigid, that I was arch-
ing my back, as though something were pulling me *into*
the lights.

IN A LONG, SINUOUS LINE, my experience in the dark zone
of Martens Cave can be tracked to a ritual performed by a
hunter-gatherer society in southern Africa known as the
!Kung San. The San, as we recall, are genetically the world's
oldest preserved population, as well as one of the most isolated: anthropologists often study their rituals to gain in-
sight into the beliefs of ancient hunter-gatherer societies
long lost to history. One such ritual is the "trance dance." It
begins at night around a fire, where the tribe claps out intricate rhythms of sacred songs. A shaman from the tribe
begins to stomp his feet to the rhythm: at first, a casual
dance, with children bouncing alongside, but as the hours
pass, he dances more and more vigorously. Dawn approaches, the dancer sweats, and hyperventilates, and grows

feverish: finally, he staggers and collapses, shaking and twitching on the ground in a semiconscious state, eyes rolled back to the whites.

In this state, the San say, the shaman undergoes a temporary death. His spirit departs from his body and travels to a spiritual otherworld, a journey that begins with a descent into a subterranean realm. One San shaman, Diä!kwain, described his spirit as "traveling far through the earth and emerging at another place." In the otherworld, the shaman's spirit performs tasks. He may escort the spirits of the deceased into the afterlife, or summon ancestral spirits to bring rain, or control the movement of game animals. Upon emerging from the trance, the shaman reports his findings in the underworld to the rest of the tribe.

Scholars of religion call this condition of semiconsciousness an "ecstatic state," from the Greek *ekstasis,* meaning "to step outside of oneself"; psychologists call it an "altered state of consciousness." Researchers have long known that consciousness exists on a many-layered spectrum. In 1902, William James wrote, "Our normal waking consciousness, rational consciousness, as we call it, is but one special type of consciousness, whilst all about it, parted from it by the filmiest of screens, there lie potential forms of consciousness entirely different."

Researchers divide the spectrum into a sequence of stages between everyday waking consciousness and the unconscious dreaming state. When we move along this trajectory, as we do each night when we fall asleep, we become more and more detached from external stimuli—the sights, sounds, and smells of our immediate surroundings—and turn our focus inward, onto the unconscious. Our thoughts loosen, drift from the linear logic of waking consciousness, until we are floating in the liquid of dreams.

But this trajectory can be manipulated. If our brain's electrochemistry is altered in particular ways, we may artificially induce the inward-focused state, enabling us to enter a dream state even as we are awake, as in the case of the dancing San shaman. In semiconsciousness, we experience visions and hear voices that are searingly vivid. According to University of Melbourne anthropologist Lynne Hume, who studies altered states of consciousness in traditional cultures, we block the "logical, rational thought processes and become open to non-mundane experiences." In this state, we access "knowledge that is different from that achieved via intellect and reason."

In the modern West, we associate altered states with drug use or pathology, a condition to be medicated or treated in the psych ward. But in the pre-modern world, altered states of consciousness were central to—perhaps the very heart of—religious experience. Virtually every person walking the planet today is descended from people who believed that a trance state was a way to summon divine power, to access the spirit world. According to a 1966 survey of 488 traditional cultures around the world by anthropologist Erika Bourguignon, 437—ninety percent—practiced some form of trance ritual. (This figure is now deemed to be closer to a hundred percent, as Bourguignon discounted many southern African cultures that have since been shown to employ rituals of altered states.) Altered states are still widely found in religious practice: voodoo priests in Haiti speak in tongues; Sufi mystics dance themselves into trances; congregants in Pentecostalist churches are taken with spirits and experience epileptic-like fits.

The details of the rituals vary from culture to culture, but the beliefs follow the same basic template: as the shaman or priest enters a trance, their spirit departs the body and travels to the otherworld, where they gain access to mystical power and superhuman wisdom, before returning to the earthly plane, and re-entering the body. The various methods of inducing trance—"techniques of ecstasy," as Mircea Eliade called them—re-create the neurological experience of entering dream consciousness, by blocking or isolating the flow of sensory input, creating a "mentally induced anesthesia of corporeal senses." To this end, people would ingest psychotropic drugs, or fast, or dance vigorously, or sing themselves to exhaustion, or play hypnotic drum music.

Or they would go underground. The dark zone of a cave has long been the ideal mise-en-scène for inducing altered states of consciousness. Celtic seers cloistered themselves in caves before delivering prophecies. Tibetan monks and lamas meditate in mountain caverns. The shamans of the Shoshone and the Lakota, among many other Native American tribes, remove to underground caverns to embark on "vision quests"; as do the mystics of the Wolof culture in Senegal, and Murut shamans in Malaysia. In ancient Greece and Rome, the oracles always delivered prophecies from underground: the famous Sibyl of Cumae, who led Aeneas into Hades, dwelled deep inside a cave, where she entered trance states and uttered sacred riddles. The rituals conducted around the all-powerful oracle at Delphi, too, were centered around a cave. (Indeed, the word *Delphi* is said to derive from the word *delphos,* meaning "hollow.") When Pythagoras enclosed himself in caves, he was inducing a form of altered state, in order to venture beyond the earthly world.

It is impossible to overstate just how deep and wide this tradition runs. Mohammad receives his first messages from Allah in the cave of Hira in Saudi Arabia. The sage Rabbi Simeon ben Yahai spends twelve years in a cave studying the torah; when he finally emerges, he burns people with his gaze. Elijah first hears the voice of God in a cave, as does Saint John, who receives the visions that become the Book of Revelation while sitting in the darkness of a cave on Patmos. When Moses asks to see the face of God, he is placed in a "hollow of the rock." Today, if you join a group of Holy Land tourists on a visit to the summit of Mount Sinai, you will be shown the cave where Moses received the Ten Commandments.

In his "Allegory of the Cave," Plato tells us that the path to wisdom is up, that reason and logic lie overhead, in the light-filled heavens. It is when the prisoner departs from the darkness and murk of the cave and ascends to the surface that reality becomes clear to him. What Plato does not say is that the world contains another kind of wisdom: an older, earthier wisdom that runs deeper than logic and reason. To access this darker wisdom, the route is down, deeper into the cave. We go into the dark to touch the divine, the mystical, the obscure.

Go into almost any cave in the world, climb down past

the twilight zone. In the dark, you will encounter remnants of ancient religious rites: burials laden with grave goods; sacred paintings traced on the walls; stone altars bearing marks of ancient fires; bone flutes to play ceremonial songs; the frenzied footprints of ritual dances; and the skeletons of animal and human sacrifices.

AFTER SOME TIME, the orbs of light overhead flickered and began to dissolve. I let my muscles uncoil, allowing the tension to seep from my neck, letting my body flatten on the floor of the cave. I blinked, blinked again. The darkness around me sank back into stillness. For some time, I stared ahead into the dark, reflecting on what I'd just experienced. It struck me as extraordinary, somehow, that my reactions to darkness were involuntary, reflexive, like a knee jerking beneath the tap of a rubber mallet. The pulsing lights were conjured entirely by my biological architecture, rising from my brain and nervous system, structures that are part of the basic hardwiring in the brains of every *Homo sapiens* to have walked the earth. Which is to say, the sensations I felt in the darkness of Martens Cave have been experienced by people all over the world, for hundreds of thousands of years.

Psychologists call these small hallucinations "entoptic phenomena"—from the Greek *entos-* (inside) and *-op* (seeing), referring to their origin from within the brain and visual-processing system. The glowing orbs of lights I saw in Martens Cave—along with other simple geometric patterns, including lines, grids, lattices, zigzags—mark the first stage of entering an altered state of consciousness. They are universal: San shamans, Tucano shamans in the Amazon, and Altai holy men in Siberia all report seeing entoptic phenomena in the early stages of entering trance states, just as they are described by subjects in Western neuroscience experiments, such as the volunteers in Donald Hebb's Project X-38.

There is a second pattern in our experience of an altered state—also universal—that brings us to the very core of our relationship with the underground landscape. The South African anthropologist David Lewis-Williams noticed it in the early 1980s, in ethnographic descriptions of shamans entering trance states. All over the world, when shamans sink into the deepest stage of an altered state, in which they undergo a ritual death and travel to the spirit world, they describe the sensation of descending through a dark hole in the ground—a vortex, or a subterranean portal. Just as the San shaman Diä!kwain recalled "traveling far through the earth," so does an Inuit shaman describe his passage into the spirit world as "traveling a road right down through the earth," where he "almost glides as if falling through a tube fitted to his body." Holy men from the Conibo tribe in Peru recount following the roots of a tree down into the earth. In the Algonkian tribe of Canada, a shaman describes "the pathway of spirits" as "a hole leading into the bowels of the earth."

The very same imagery, as it turns out, appears in contemporary psychology: as laboratory subjects enter the deepest phase of an altered state, they feel as though they are descending through a dark passage in the ground. In one study, the UCLA neuroscientist Ronald Siegel found that in fifty-eight reports of eight types of hallucinogenic imagery, the most common was the sensation of passing through a dark tunnel. It was especially prevalent in reports of so-called near-death experiences, where a person has, say, a heart attack, is pronounced dead in the ambulance, but is revived. In the wake of such episodes, a patient will frequently report sensations almost identical to those of a shaman entering a trance and undergoing a ritual death. In a study that briefly emerged as a bestselling book during the 1970s, a psychiatrist named Raymond Moody interviewed 150 people about near-death experiences. The most frequently reported sensation in all of his subjects was the feeling "of being pulled very rapidly through a dark space of some kind." He listed the various given analogues as "a cave, a well, a trough, an enclosure, a tunnel, a funnel, a vacuum, a void, a sewer." One survivor reported traveling through a portal that "just fit her body," as she said. "My hands and arms seemed to be at my sides. I went headfirst, and it was dark, dark as it could be in there. I moved through it downward."

In fact, the previous afternoon, just before I'd entered Martens Cave, Craig Hall, my host in West Virginia, had told me precisely that story. I was sitting on the bumper of my car, lacing my boots, when I asked, "Have you ever done a solo cave sit before?"

"No, not really," he said. "But sometimes I'll let the other cavers go to a different part of the cave. And I'll turn out my light and just sit alone in a chamber."

"Have you ever felt anything strange?" I asked.

"You mean, have I had visions?" he said. "No, nothing like that."

I nodded, went back to my boot laces.

"What it feels like," he said, "is when I died."

He paused.

"When I was in my twenties, I had mononucleosis and was in bed for weeks. One night—there's no other way to say it—I died. I could see myself. I could see my family. I met a kind of spiritual being. But he turned me away. Because it wasn't my time. I don't know how long I was out. When I'm in cave darkness, I feel the same way as that night." He paused. "Like I am outside of my body and I'm moving *inside* the earth and I can see everything at once."

"WE ALL HAVE A CAVE in our mind." I said the words aloud to myself, measuring how they fell in the dark. That is to say, our brain is structured in such a way that the sensation of moving beyond normal consciousness resembles the sensation of entering a cave. "Caves are the topographical equivalents to the psychic experience of the vortex and entering the nether world," wrote David Lewis-Williams in his 2002 book *The Mind in the Cave*. It is an echo that has been ringing through human culture for longer than we can possibly say. Deep in the past, our ancestors told stories of portals in the mind, where they traveled through dark mental passageways, experienced death, passed into a plane of consciousness beyond everyday reality. And they told stories about portals in the earth, where they climbed down into rocky caves, lit their way through the dark with pine torches, moving through unearthly environments unlike any surface landscape. Over time, stories of mental landscapes and physical landscapes bled together, until they became indistinguishable, until mind-portals and earth-portals were one.

As our ancestors circled the planet, every culture in the world told stories of such portals, where heroes traveled through dark passages in the earth and entered the spirit world, then returned to the surface, endowed with sacred wisdom. Just as Pythagoras traveled to Hades through the Cretan cave, so did cultural heroes all over the world from the Maya and the Celts to the ancient Norsemen and the Navajo. Even Jesus Christ descended to the underworld by way of the subterranean dark zone. In the "Harrowing of Hell" in the Gospel of Nicodemus, one of the apocryphal books that was initially excised from the Bible, Jesus is enclosed in his rocky tomb—which, we recall, is a cave with a stone rolled over the entrance. From within the darkness of the cave, Jesus departs from his earthly body and descends into Hell, to "the lowest places," where he preaches to the dead, and frees wrongly imprisoned souls. It is from the underworld that Jesus is resurrected and ascends to Heaven.

One of humanity's oldest recorded stories, *The Epic of Gilgamesh*, which was inscribed by the Mesopotamians four thousand years ago on a clay tablet, is a tale of descent. Gilgamesh is traveling to the otherworld to find the secret of eternal life. To reach the beyond, Gilgamesh, whose name translates to "He Who Sees the Deep," must travel through a long, dark tunnel.

. . . Ever downward
through the deep darkness the tunnel leads
All will be pitch black before and behind
all will be pitch black to either side.

It is an ambiguous tunnel, lacking any concrete detail, leaving us unsure if he's moving through a dark passage in the earth or a dark passage in the mind.

Each time we peer into the mouth of a cave, or a tunnel, or any other hole in the ground, we feel a flicker of recognition: we've seen this space in our dreams, on the edges of consciousness. When we pass through this portal, we know that we're leaving behind the clarity of the surface world, withdrawing from the linearity and logic of ordinary consciousness, slipping into the fluidity of the unconscious. We are Michel Siffre, seeking to alter biocycles in the dark zone, or we are Pythagoras, communing with ancestral spirits: in either case, we are stepping outside the gyre of

ordinary reality, and edging closer to whatever lies beyond the margins of the world.

In my final hours in Martens Cave, I lay in the dark humming to myself, feeling the invisible contours of the cave manifest in the echo of my voice. As I felt myself growing restless, I took off my boots, rose to my feet, and in my blind state, I started taking small, careful steps around my camp. At first, I shuffle-stepped, wriggling my stock-inged toes exploratorily over the cave floor, feeling for boulders, to make sure I didn't trip. I made a lap around my camp, then started another lap, retracing the path, my steps now less timid, my feet now lifting slightly off the ground. Then another revolution, and another, until I was taking full strides, tracking circles in the dark.

I emerged from the cave just before seven in the eve-ning, having completed the twenty-four hours. I stood on the edge of the ravine, blinking in the light. As my pupils shrank back into points, I watched, in the words of the poet Mark Strand, "as the world assembles itself once more." A caver once told me that being in a cave is like being dead, but also like being unborn, and now I felt both as though I were returning to the world from beyond, and entering the world for the first time. Finally I hoisted my bag over my shoulder and hiked out into the forest, where I felt a re-stored gratitude for the faintest things—light, air, warmth, clarity. I felt Strand's ecstatic final stanza echoing in the quiet chambers of my mind:

> *Thank you, faithful things!*
> *Thank you, world!*

To know that the city is still there,
that the woods are still there,
and the houses, and the humming of traffic,
and the slow cows grazing in the field;
that the earth continues to turn
and time hasn't stopped,
that we come back whole
to suck the sweet marrow of day.

CHAPTER

9

THE CULT

The city's gods, according to some people,
live in the depths, in the black lake that feeds
the underground streams.

—ITALO CALVINO, *Invisible Cities*

The Yucatán Peninsula of Mexico may well be the most perforated place on the planet. It's a land so sponged with caves and potholes, crevices and pits, you must walk with your eyes steady to the ground, or you may tumble down into the earth. In the way people of the Arctic dream at night of glaciers, and bedouins of desert dunes, the inhabitants of the Yucatán have long found their quiet thoughts occupied by caves.

On the afternoon of September 15, 1959, a young man named José Humberto Gómez picked his way into one such hollow, a small jungle cave known as Balankanché. The cave was hidden in the forest a few miles from Chichén Itzá, the ancient Maya city of soaring pyramids and graceful stone courtyards. Originally recorded by archaeologists in the early 1900s, Balankanché was not a celebrated cave, or even a noteworthy one. Its few dank chambers contained a smattering of ancient Maya potsherds, a lot of bat guano, and little else.

Humberto—a wiry, bright-eyed man in his early twenties—worked as a guide for visitors to the ancient cities. He had been coming to the forest since he was a little boy, to be with his grandmother, who helped run a hotel in the forest called the Mayaland Hotel. As a boy, Humberto would ride a horse into the forest each morning, following paths between jungle villages such as Xcalacoop, where the residents were Maya, descendants of the people who'd built the stone cities. He'd spend entire days clambering over jungle-choked ruins, most of them still uncharted by archaeologists, then return to report his findings to his grandmother at the Mayaland. When Humberto turned thirteen, the hotel's head gardener, an older Maya man named Bel Tun, who knew every groove in every corner of the forest, told him of a cave hidden in the jungle. People had stopped visiting the cave many years before, Bel Tun said, but perhaps Humberto would find something interesting there.

The first time Humberto entered Balankanché, he lit his way with candles he'd collected from Christmas posadas at the hotel. He lit one candle, stuck it in the mud, then lit another candle, and another, following the flickering trail down into the dark zone. From that day on, Humberto felt a magnetic pull from Balankanché that he didn't quite understand. Even though the cave contained nothing special, he came back again and again. He'd dig in the dirt, searching for artifacts left by ancient visitors, or he'd just sit and feel the darkness press on his skin. Sometimes he'd bring friends down, but they never seemed to feel what he did, and Humberto would be left wondering what was wrong with them. Humberto went to university to study anthropology, but then left, preferring life beyond the classroom—he missed roaming in the forest, searching for ruins, and visiting the cave, which he had come to know as intimately as his own home.

On that afternoon in 1959, Humberto was down at the back of a passageway he'd visited hundreds of times, when he saw something he'd never noticed before: an oddly colored patch in the stone, half-obscured by mud. When he scraped away the mud, he was stunned to find a wall of bricks, made in precisely the style of masonry he knew from the ancient cities. With his knife, he chopped around the bricks until he broke through, revealing a tunnel receding into the dark. With his heart thumping in his chest, he crawled ahead. Finally, he emerged into a large echoing chamber, where he froze.

At the center of the room was a floor-to-ceiling stone column that branched out at the top and bottom like the boughs and roots of a tree. At the foot of the pillar, on the slimy cave floor, Humberto passed his light beam over a ceramic pot. Next to that, another pot. Then dozens of them: pots, incense burners, urns, all painted in dazzling colors and carved with the faces of gods. Water dripped from the upper branches of the pillar, falling in and around the pots. Humberto stood nailed to the floor, listening to the percussion of the drops in the dark. He was the first person to set foot in this chamber in twelve hundred years.

News of the discovery trickled through the forest. A few days later, as a crew of American archaeologists were heading into the cave, a man named Romualdo Ho'il appeared at the entrance. Ho'il was the shaman of the village of Xcalacoop and studied the archaeologists with a grave eye. The ceramic pots, he explained, were offerings given by his ancestors to the lords of Xibalba, the Maya underworld. By entering this sealed chamber, they had awakened forces powerful beyond their understanding—he would have to cleanse the space.

Ho'il returned with a group of men from the village, who all filed into the cave and gathered around the pillar. Humberto and the archaeologists, who likewise needed to be cleansed, were ordered to stay. The ceremony lasted twenty-nine hours: Ho'il sacrificed thirteen chickens and one turkey, lit copal incense and black candles of wild bee honeycomb, and drank great quantities of *balche*, a sacred wine made from fermented tree bark and honey. As the hours passed, the oxygen in the chamber diminished and the darkness grew heavy with smoke, until it was almost impossible to breathe. The shaman emitted guttural sounds, imitating the voice of a jaguar, while other men chirped like frogs. They danced and prayed and sang, their voices

rising in a wild chorus. When the ceremony came to an end, the supplicants proceeded to the surface, where they emerged into a thunderstorm, where rain whipped down from a blackened sky.

WHEN I FIRST READ about the discovery of Balankanché, I saw in Humberto's cave excursions a reflection of my own childhood trips into the tunnel under Providence. The landscapes were different—Mexican jungle; shaded New England street—but both of us had been young boys who'd developed an intimate connection with an otherwise unremarkable underground space. Even Humberto's ancient ceramic pots, with water dripping from the ceiling, felt like a shadow of the buckets I'd found in the tunnel, where water drummed down from above, sending an echo through the dark. After so many years of exploring my relationship to the underground, I wondered if I might talk to Humberto about his discovery at Balankanché, to hear the ways in which it had impacted his life.

But Humberto soon drifted from my mind as I learned that his discovery was just a single point in a vast tradition of cave worship that pervaded Maya culture. The Maya territory—ranging from the Yucatán, down through Belize, Guatemala, Honduras, and El Salvador—was riddled with an entire galaxy of caves, from grand limestone caverns to water-filled sinkholes called *cenotes,* and every single one of them was believed to be a spiritual portal into the underworld of Xibalba. In the years since Humberto's discovery, every time an archaeologist dropped underground, and entered the dark zone of a cave, they found ancient offerings. Sometimes just a few ceramic pots or pieces of jade or stingray spines, but other times, they found remains of sacrificed deer, jaguars, crocodiles, as well as the skeletons of humans. In some caves, they discovered entire paved roads and brick-and-mortar temples constructed in the dark. The ancient Maya often risked their lives to deliver these offerings, swimming down subterranean rivers, climbing precipitous cliffs, and lowering themselves down into perilously tight hollows.

As I continued to speak to archaeologists working in the jungle of Mesoamerica, I found unfolding before me an entire culture of underground obsessives, a people whose very existence depended on their relationship to caves. The Maya built their cities near caves, chiseled sculptures of caves into the walls of temples, painted images of caves on ceramic pots. They performed dances and sang songs about caves. In the intricate hieroglyphic script of the Maya, one of the most common glyphs was the symbol for "cave"—*ch'en*. The founding epic myth of the Maya, the *Popol Vuh*, tells the story of two brothers, the Hero Twins, descending into the underworld of Xibalba. Here were people who worshipped before caves in the day, told stories of caves in the evening, and dreamed of caves at night.

"It's underworld mecca down here," an archaeologist told me one afternoon over the phone. Holley Moyes was from the University of California, Merced, and had spent two decades in the jungle, crawling through bat guano, knocking her helmet against rocky ceilings, documenting the cave cult of the Maya. She was known among her colleagues as the Queen of the Dark Zone.

Her research, I learned, went far beyond the Maya. For years, Holley told me, she had been digging through ethnographic and archaeological studies on the role of caves in traditional cultures around the world. In 2012, she published a book, *Sacred Darkness: A Global Perspective on the Ritual Use of Caves,* in which she compiled research from archaeologists and anthropologists on the relationship to caves in over fifty cultures, on six continents, ranging from the present day back to the Paleolithic Age, some 100,000 years ago. She presented evidence of a virtually universal tradition of subterranean religious practice, carried out in every corner of the planet, at every point in history.

It was a déjà vu moment: the feeling of brushing shoulders with a stranger on the street who inexplicably feels like an old friend. I explained that I had spent years traveling all over the world, documenting our connection to the underground: the ways we are repelled by the dark, and yet compelled by mysterious impulses to venture down into the earth. Holley went quiet on the phone, then laughed out loud.

"Well, you better come down to the jungle," she said. "We have a lot to talk about."

ON A SWIRLING, SUNLESS afternoon in August, as a tropical storm gathered in the Gulf of Mexico, Holley picked me up at the airport in Belize City. She was in her mid-fifties, with auburn hair that skimmed her shoulders, and lively, expressive eyes.

"Hope you don't mind a little dirt," she said, nodding to her Jeep, which looked like it had been hauled up by its bumper and lowered into a vat of mud.

From the coast, we headed inland toward San Ignacio, a small town nestled in rain forest, where Holley kept her research station. It was the rainy season and around every bend, we crossed brown, frothy rivers, kicking up waves of orange mud. Two decades of fieldwork in Belize were enough that Holley knew the region like a native, but not so much that she was inured to the ferocities of the landscape. Holley told stories of defending excavation sites from gun-wielding looters; sniffing out jaguar urine in the forest; negotiating with local shamans for access to sacred places; digging trucks out of flooded forest roads; dodging snakes and vampire bats and scorpions and "assassin beetles," which carried the deadly Chagas disease.

Deeper into the forest, the air grew cool and fresh, as we weaved through mammoth, emerald green hills. I scanned the scenery, knowing that hidden everywhere in the trees were vestiges of ancient Maya settlements. During the height of the Maya civilization—from around A.D. 250 to A.D. 950—this was home to hundreds of thousands of people, living in some of the grandest cities in the world at the time. Cities like Tikal, Copan, and Palenque thrived on terraced farms cut into the faces of mountains; the land flourished during the rainy season, and the Maya constructed networks of cisterns to conserve water for the dry season. For centuries, they lived in opulence. They became great mathematicians, and produced wondrous artworks. They erected lordly pyramids that rose above the tree horizon, built ornate stone temples, and installed colossal stelae, or stone pillars, engraved with the histories of their divine kings. But like all civilizations, the Maya fell. Around the ninth century A.D., a terrible drought took hold of Meso-america. The rain they depended on to grow crops ceased, and the cities were unable to support their populations: starvation set in, millions perished.

"As things got more and more desperate," Holley told me as we drove, "they became obsessed with caves. Every-thing became about going into the dark zone." The follow-ing morning, we planned to visit a cave called Actun Tunichil Muknal, the "Cave of the Crystal Sepulcher," which was the first cave Holley had studied in Belize, the cave where she'd seen the very first evidence of the ancient dark-zone cult.

Under a low, mercury-colored sky, Holley and I trekked through the Tapir Mountain Nature Reserve, about fifty miles from San Ignacio. It was dense, prelapsarian jungle, where the air was thick and soupy and everything smelled of moss. We scrambled over giant buttressed roots and waded through rivers that flowed up around our waists. Iguanas skittered through the undergrowth, and tanagers and toucans chirruped in the trees overhead. In the distance, we could hear the percussive *whoop* of howler monkeys. Before long, we broke through a wall of vegetation and I found myself staring down into a gaping cave mouth: a smooth-sided, hourglass-shaped aperture, with vines dangling over the lip. Out of the entrance flowed a river, tumbling quietly between mossy boulders.

"The Maya depicted caves in their artwork as the mouths of a monster," said Holley, pointing to the stalactites jutting down from the upper lip of the entrance. "You can even see the teeth."

After a beat, she added, "It looks a heck of a lot like a vagina, too."

From a large boulder, we leaped into a pool of warm, translucent green water as a pointillistic cloud of minnows darted beneath us. We breaststroked across the cave threshold, sunlight fading to twilight, then dropping off into pitch-darkness. We worked against the current, scrambling up over slimy boulders, toppling down into whirlpools, twisting and squeezing through keyhole slots, as the river water sputtered and gushed around us. Holley—who'd been one of the first archaeologists to study the cave

after it was located by amateur British cavers in 1986—
maneuvered through the noisy choke of boulders on pure
muscle memory, as though following the steps to an old
choreographed dance.

We emerged into perfect stillness: a colossal, pitch-dark
hallway, where our headlamp beams crisscrossed back and
forth overhead, like city spotlights. I looked for bats swoop-
ing above, or hanging from stone, but the ceiling was too
high to see. As we swam, water plop-plop-plopped into the
river all around us.

About a half mile deep, we drifted over to the river's
bank and hoisted ourselves up onto the stony ledge. Holley
instructed me to remove my boots: in stockinged feet, we
tiptoed into the cave's central chamber, which was ringed
with glittering stalactites and stalagmites and massive
floor-to-ceiling pillars, as thick as tree trunks.

As I swept my headlamp beam through the chamber,
my breath caught. Spread across the floor were hundreds of
ancient ceramic pots, painted in black and dazzling orange.
Some were as large as beach balls, and had been cemented
in place with centuries' worth of calcite growth. Scattered
among them were stone tools, shards of jade and obsidian,
and small statues of animals, including a stone whistle in
the shape of a dog.

"All of these artifacts date to the ninth century," Holley
said. "From the time of the drought."

Up an iron ladder installed in the cave wall, Holley
guided me to a narrow alcove above the chamber. "There
she is," she said, crouching low on the ledge. We were looking at a human skeleton—a twenty-year-old woman.

"We call her the Crystal Maiden," she said. I swallowed hard. She was lying on her back, her legs spread-eagled and arms akimbo, her ribs encrusted in calcite, giving the bones a haunting crystalline sparkle. The skeleton was perfectly articulated, except for the jaw, which was jarred open in a tilted, frozen smile.

"Notice that there are no grave goods here," Holley said flatly. "This was not a burial."

The Crystal Maiden was not alone. Back down on the floor of the central chamber, the ground was littered with skeletons—fourteen in all. At the foot of a hulking stalagmite were the remains of two young men, both beheaded, their skeletons half-dismantled and coated in calcite. Nearby, the skeleton of a man in his forties, his temple bludgeoned. We crouched over the victims, visiting them one by one, including the remains of an infant: a pile of tiny bones, tucked away in a dark crevice.

All of these people had been sacrificed as offerings to Xibalba.

"Xibalba," said Holley as we crouched in the dark, "was different from the way we think of Hell."

For the Maya, the "Place of Awe," as Xibalba was translated, was not abstract: it was a tangible, geographical place, a place you could point to on a map. On a hike through the forest, you could smell Xibalba, you could hear its rumbles and its echoes, you could feel a breeze blowing up from its depths. And if you climbed down through the rocky aperture of a *cenote*, or into the mouth of a cave, and slipped over the threshold of the dark zone, you were stepping *inside* Xibalba. You were leaving behind the earthly world and entering an entirely separate realm, where you would come face-to-face with spirits, deities, and beings of mercurial power.

The Maya connection to Xibalba was visceral, peculiar, and riven with ambiguities. In the *Popol Vuh*, when the Hero Twins—Hunahpu and Xbalanque—descend into Xibalba, they navigate a warren of horrifying compartments: one chamber that rages with fire, another spiked with daggers, another full of prowling jaguars. At every step, the Hero Twins battle the Lords of Xibalba, a gang of repulsive creatures with names like Seven Death, Pus Master, Jaundice Master, Blood Gatherer, and Stab Master, who spend their days spreading disease and desolation in the surface world. And yet, as forbidding as the underworld was, the Maya depended upon Xibalba: they couldn't live without it. Living alongside the Xibalbans was the god of rain, Chaak.

He was an impetuous and wild-eyed deity, who brandished lightning bolts and unleashed thunderclaps over the forest—but he provided rain, without which the Maya could not survive.

For centuries, to keep Chaak satisfied and ensure that he continued to bring rain, the Maya would leave gifts for the god at the mouths of caves. They would crawl underground—staying always within the reach of light, keeping a safe distance from the dark zone—and leave offerings of ceramic pots and shells of the sacred *jute* snail.

For centuries Chaak was content with the gifts: each year, as the dry season ended, and the planting season began, the god delivered rain, the crops grew, and the Maya thrived.

But then, all at once, Chaak abandoned them. During the eighth and ninth centuries, for reasons the Maya could not fathom, the god retreated into the hidden recesses of the underworld and refused to emerge. Rain stopped falling, crops in the terraced hills withered. For a while, the Maya persisted with the same ceremonies that had brought such prosperity to their ancestors, delivering pottery and *jute* shells to cave mouths—but Chaak did not stir. They tried leaving more generous gifts, delivering larger mounds of pottery and shells, sometimes even leaving a freshly sacrificed animal at cave mouths—still, no answer from Chaak. Soon they were desperate: children in the cities were starving, people spoke of abandoning their homes and wandering north. They had one last hope to please Chaak, to regain his favor: they would journey into Xibalba, pass into the dark to meet the god in his own territory.

Twelve hundred years before Holley and I, a small procession of Maya people waded through the mouth of Actun Tunichil Muknal. They drifted to the far edge of the range of diffuse light; after a moment of quavering hesitation, they pushed ahead, as though stepping off the edge of a cliff. They were priests, wearing feathered regalia. They were thin and haggard, their faces drawn. They hoisted ceramic pots full of maize, carried grinding stones, and fragrant copal to burn. One carried an obsidian blade sheathed on his hip. At the center of the group was a twenty-year-old woman who walked in silence, the river flowing around her neck.

They wended slowly upstream, walking in single file, their flaming pine torches casting a smoky glow through the dark. No one spoke, and they moved warily, every step made in trepidation. Like everyone living in the forest, they had heard and told stories of Xibalba since they were children, but this was different. They passed through stifling darkness, touching their fingers to sweating stone walls, watching the shadows of rocky spires trembling in torchlight. They glimpsed albino fish darting through the water, heard the riffle of bat wings overhead. When a stone splashed into the river ahead, sending an echo through the darkness, they all went tense. But they pushed on: if anything would coax the god of rain out of hiding, it would be this journey into the dark.

A half mile into the cave, the priests rose out of the river, processing into the central chamber, their torches setting light to looming stalactites and stalagmites. They laid out their gifts for Chaak, hefting the pots down from their shoulders, spilling maize over the stone. As they prepared the ceremony, they lit the sacred copal. With the fragrant smoke twisting up into the chamber, they began to chant an invocation to Chaak, lifting their arms in the dark, as they gathered around the young woman. The priest unsheathed his obsidian blade and raised it into the air: as their voices swelled, and rang up among the stalactites, he brought it swiftly down.

AFTER PULLING ON OUR BOOTS, Holley and I climbed gingerly down to the rocky bank, and eased back into the river. We began a slow breaststroke downstream, as water dripped into the river around us. The ceremonies conducted in the dark zone of Actun Tunichil Muknal, she said, were not unique. Over the past decade, archaeologists had been recording the dates of offerings discovered in the dark zones of caves throughout the Maya territory. Virtually every artifact—every ceramic pot, every stone tool, every bone from every human sacrifice—dated to the time of the drought. At a cave called Chechem Ha, just a day's walk from Actun Tunichil Muknal, Holley had found a stone stela placed upright in the dark zone, surrounded by ceramic pots and traces of fires, all dating from the ninth century. Most recently she'd been excavating another nearby cavern, this one called Las Cuevas, where she'd discovered an elaborate assemblage of ceremonial platforms and stairways, all built by the Maya during the time of the drought. "It's not just around here—it's *everywhere*," she said. Even the offerings discovered by Humberto in the concealed chamber of Balankanché dated to the ninth century. The ceramic pots were carved with the wrenched face of the rain god. "We're talking about an enormous collective ritual," she said, "conducted all throughout the forest."

As we drifted down the river, both of us now quiet, water lapping at our exposed shoulders, I turned Holley's words over in my head. A scene slowly materialized before me, first in shadow and silhouette, then sharpening, until I could see it in expansive detail, an extraordinary and haunting tableau. I saw thousands of pilgrims, people in their most desperate hour, scattered throughout the Maya land, all of them moving, as though part of a giant single body. I saw them hiking through the forest, like shadows in the trees, until they arrived at the mouths of a thousand different caves. They crouched for a moment in the twilight zone, then, with a collective intake of breath, they all struck ahead into darkness. Deep underground, the pilgrims danced and prayed and sang, their disparate voices all rising in the dark as a single voice. They delivered gifts, laid out jade and obsidian offerings, and they performed sacrifices, disemboweling animals, and spilling the blood of men, women, and children over the damp stone floors. Beyond the barbaric violence of the scene, beyond the fact of its being an apocalyptic vision, I found myself marveling at this collective ritual as a display of prodigious faith and devotion. Here was an entire civilization who, in their most desperate moment, as death closed in on them, called upon the power of the underground world. A people who ardently believed that these hidden chambers, in their eternal darkness and rumbling echoes, were sacred and magical, that they possessed the power to reshape reality.

I drifted down the river, thinking of all the ancient processions that had waded down this corridor before us, all of the people who had stepped warily through this very darkness, listening to the same echoes ripple along the walls. As I allowed my thoughts to soften, something curious happened, where the temperatures of the water, the air, and my skin all began to converge, until all three forms of matter became indistinguishable from one another. In this peculiar state, I found myself surrendering to the current, letting myself go slack, as though the edges of my body were dissolving, and I could no longer discern where my skin ended and the cave began.

THAT EVENING, HOLLEY AND I sat on a picnic table on the back porch of her research base. The night air was soggy, and a guttering citronella candle cast orange light on our faces. We spoke about the day's excursion into Actun Tunichil Muknal, ruminating on what it meant to follow in the footsteps of the Maya, what had drawn them into the dark, and what drew us.

"We have a need for the sacred," said Holley, as she took a long sip of water. "We all have a desire to seek out God, or the gods, or spirits, or magic—whatever you choose to call it. It's innate to being human."

Ours has always been a spiritual species. "Man is by his constitution a religious animal," wrote Edmund Burke in the eighteenth century. Since then, neither anthropology nor history has ever known a human society that did not observe some form of religion. Today, few evolutionary biologists, theologians, or cognitive scientists would deny that spiritual impulses are hardwired, engraved in human nature. From the emergence of *Homo sapiens*, hundreds of thousands of years ago, we have possessed a brain with a powerful buzzing neocortex that enabled us to shape thoughts that other members of the animal kingdom could not. We pondered our own existence, carried ideas beyond our conceptual grasp, and formed relationships with dimensions outside of what we could touch or see. As we moved through the planet, we dedicated vast amounts of energy and resources to religion: we composed lyrical prayers and devised ceremonial dances to honor deities and spirits, we built tombs for our ancestors, we constructed temples with spires that reached to the heavens and carved crypts that reached down into the earth. The desire to connect with something larger than ourselves, wrote British scholar of religion Karen Armstrong, may well be "the defining characteristic of humanity."

It was *this* impulse that first drew our ancestors underground. In the dimmest reaches of prehistory, our ancestors climbed into the darkness of caves on quests to communicate with the spirit world. In the cosmologies of ancient cultures all over the world, the cave environment *was* the spiritual dimension of reality. To go underground was to step bodily *inside* the otherworld—"the world behind this one that we see with our eyes," as the San called it. Just as the Maya did in Actun Tunichil Muknal, our ancestors everywhere conducted sacred rites in the dark to summon supernatural powers.

"It's astonishing how far back this tradition goes," said Holley. She told me of a cave in the Atapuerca Mountains in northern Spain, where, at the very deepest point in the dark zone, at the bottom of a forty-foot-deep vertical shaft, a team of archaeologists discovered a jumble of human bones. The Sima de Los Huesos, or the "Chasm of Bones," as the site became known, contained the remains of twenty archaic humans, dating to between 430,000 and 600,000 years ago, long before modern *Homo sapiens* even existed. Discovered amidst the bones was a hand-axe of glittering red quartzite—a rare stone, imported from a great distance, signifying that it was special—which archaeologists called Excalibur. Many researchers believe it is the first evidence of religious behavior: an archaic dark-zone ritual, honoring a journey into the afterlife.

In the modern West, of course, we no longer connect to the world in this way. We are a post-Enlightenment, industrial society—a people of science and technology—whose perception of reality, for the most part, is grounded in reason and rationality. Over the last few hundred years, since the first writings of Descartes and Spinoza and other Enlightenment philosophers, Western culture has grown steadily more secular. Where religious faith consumed the entirety of existence for our pre-modern ancestors, today we see religion occupying a distinct sphere, something outside the prevailing doctrine. "Modern man," Eliade wrote, "has forgotten religion."

When we drop through the mouth of a cave, we do not believe, in any rational horizon of our mind, that we are departing the earthly realm and entering the spirit world. And yet, we fall directly into step with those who *did* believe. We follow precisely the same footholds as our ancestors, we stoop and crawl and twist our bodies at the same angles, hear our voices echo, and feel our breath against the stone walls in the same way. On our way into the dark, we perform an unwitting shadow-performance of the old rituals, sometimes following the ancient choreography down to the last gesture. Having the same bodies and minds as our ancestors, we undergo the same sensory experiences, which are just as bewildering and unsettling and thrilling to us today as they were in antiquity. In our rational mind—according to the physical laws honed by Western scientists over centuries—we attribute these sensations to shifts in biological rhythms, to activations or suppressions of various portions of our nervous system. And yet, in the deep strata of our consciousness, we feel something that shivers beneath rationality. "There's no question that when we're in cave darkness," said Holley, "something in us shifts. We're able to confront ourselves and engage with the world in ways we never do otherwise."

In the evolution of religion, Robert Bellah wrote, "nothing is ever lost." Over the course of history, even as we have accumulated new philosophies and creeds, the elementary structures of our ancestors' beliefs have never quite disappeared, remaining always intact in our core, however deeply buried. Our connection to caves may well be our most universal, most deeply inscribed, perhaps our *original* religious tradition—which is to say, it casts a long shadow. However modern or civilized or enlightened we may consider ourselves, when we climb down into a cave, we feel something primal stir within us. We slip into a kind of ancestral muscle memory, revert to a more intuitive, animal mode: a handful of centuries of rationality, science, and empiricism are promptly swept beneath hundreds of thousands of years of instinct and evolutionary conditioning. In cave darkness, wrote Seneca, you cannot help but "feel your soul seized by religious apprehension." When even the most rational, most materialist, hardest-line atheist climbs into a subterranean dark zone, you will hear them drop their voice to a whisper—somewhere in their unconscious, they feel awe and immensity and mystery, and they recognize this as a holy place. Today we may not perform sacred rites in a cave dark zone, we may no longer know the ceremonial prayers once chanted there, but we still carry their echoes deep in our minds—the old cosmology holds firm within us. "We find ourselves in the presence of a form that guides and encloses our earliest dreams," as Bachelard wrote.

"All of this doesn't just disappear overnight," said Holley, a smile coming across her face in the dark. We no longer speak of the firmament, or of the celestial spheres, as our ancestors did in antiquity, but, as the philosopher Henri Lefebvre writes, we have not shed our belief in the underground as a potent place, "filled with magic-religious entities, with deities malevolent or benevolent, male or female, linked to the earth or to the subterranean (the dead), and all subject to formalisms of rite and ritual." Every year in southwestern France, six million Christians make a pilgrimage to Lourdes, where they follow a procession into the shadows of a small cave where a young woman was visited by an apparition of the Virgin Mary. Thousands of pilgrims each year visit Station Island in Lough Derg in Ireland, in order to walk around the place where God revealed a cave to Saint Patrick. In almost any church in Europe, directly beneath the pews where people genuflect during mass, there lies a secret chamber—hidden, but intact—where in antiquity the mysteries of the earth were celebrated.

Over hundreds of thousands of years, our vivid and perplexing connection to the underground has not diminished—and it never will. We will always feel a quiet glow emanating from the world's buried places: it may be forbidding or enchanting, but we will never look away. George Steiner wrote of a hidden "transcendent presence in the fabric of the world": the underground world *is* that presence. Just like our ancestors before us, we will always be drawn underground by a quiet desire to reach beyond the mundane and ordered reality, to touch upon something greater than ourselves. The Paleolithic hunter-gatherer crawling by torchlight into the depths of the cave, the urban explorer in Paris roaming the catacombs, the pedestrian in New York lingering over the open manhole in the street: at a deep root, all are animated by the same basic longing.

After saying good night to Holley, I climbed into my bunk bed at the research base, where I lay awake for some time. As I listened at the window to the sigh of the breeze coming down from the hills, my mind slowly turned over. I saw that all of the underground devotees I'd explored with over the years, or admired from across history, were, in one form or another, seekers of transcendence. Michel Siffre, who sought to break free of his biological rhythms in the dark zone; REVS, who made clandestine artworks in the city's bowels; William Lyttle, who burrowed beneath his home as though digging to a parallel dimension; John Cleves Symmes, who pursued intraterrestrial beings; Nadar, who captured images of the invisible layers of Paris; Steve Duncan, who walked the paths of ancient streams through quiet darkness under the city. All of them climbed underground in search of mystery, seeking contact with something beyond the immediate horizons of reality. I fell asleep that night thinking of Hermes, forefather of all of these seekers, who swooped openly between this world and the otherworld, who could see the unseen.

———

OUT OF BELIZE, I followed a long, rutted path north—by night bus, then lurching minivan, then a station wagon piloted by an old man named Jorge—across the Mexican border, until I arrived in the cave-ridden land of the Yuca-

tán, where, one afternoon, under a quavering sun, I came to the mouth of the cave of Balankanché, and found myself sitting across from Humberto. He was in his seventies now, but still looked much like a photograph I'd found of him crawling through the cave as a young man: narrow-shouldered and fastidious, with hair in a perfect pompadour swoop.

"I spent many hours as a boy sitting exactly in this spot," he said, a quiet warmth in his eyes. Behind him was the cave entrance: once hidden beneath wild ferns, now outfitted for visitors, with paved steps leading to an iron door.

I told Humberto why I'd come to see him, about my unexpected affinity for a tunnel I'd found as a boy beneath my neighborhood, where I'd discovered an altar of buckets arranged in the dark, with water drumming down from the ceiling. The tunnel, I told him, had left an imprint on my mind that I'd spent years trying to understand.

"I see," said Humberto with a quiet laugh.

"I knew this place like it was my home," he said. "On that day, to break through the wall, it was like discovering a hidden room inside of my own house. It changed many things for me."

In the Maya villages throughout the forest, Humberto told me, people began to whisper about him. A young man had traveled into the underworld, they said, where he'd unlocked a hidden chamber and made contact with powerful ancestral spirits, and then he'd returned to the surface, without a mark on him. He had been divinely chosen, they said, and was endowed with a power to see what others could not. People in the villages summoned Humberto to investigate caves in the jungle that others were too frightened to visit. *You are the one to go,* they would say. He became a kind of cave-whisperer, moving from village to village in the forest, climbing underground with his flashlight, exploring in the dark, then reemerging to report his findings to the villagers.

"I did not think of myself as going inside of the underworld," Humberto said. "I did not think I'd had a spiritual transformation. These were not my beliefs. But in certain ways ..."

He paused. "At the time of the discovery, I was a young man. I had no wife, no girlfriend. I moved between only a few places. My world was very small," he said, balling his fingers into a fist.

"When I broke through the wall, many things opened for me." He opened his fingers. "If this hidden chamber existed, there were other places to discover—it seemed many things were possible."

Following the discovery, Humberto returned to work as a guide and fell back into his old routine, only things were different. He led visitors through the ruined jungle cities, up the step pyramids, and into the quiet stone courtyards, but now he implored them to go slowly, to linger, to look more closely. There was a hidden dimension to these spaces that did not immediately present itself, an entire cosmos of history, myth, and sensation. "I wanted people to look beyond what they saw in front of them," Humberto said.

We both went quiet, sitting in the shade, listening to a

chorus of insects buzz around us. Then Humberto rose and pulled open the door to Balankanché, revealing a dark passage underground, where he motioned for me to go.

"I no longer go inside the cave," he said. The air, he explained, was dense and humid; as he'd grown older, it had become difficult for him to breathe underground.

I moved to protest, but he waved me off. "You go," he said.

Through the mouth, I headed down into the dark, stepping lightly on the slick stone floor. I passed the threshold where Humberto had broken through the wall of bricks a half century before. I went down and down, the air thickening around me, growing dense with moisture, until rolling banks of mist formed at my feet. As I emerged into the heart of the cave, I stopped at the foot of the giant pillar, which rose above me like an ancient tree, its gnarled branches spreading overhead. At the foot of the pillar, the ceramic pots were still arranged just as Humberto had first encountered them so many years before. Water dripped from the ceiling and as I stood in the dark, listening to the soft patter around the pots, I felt myself before the altar of buckets in the tunnel under Providence. I thought of the bolt that ran through me that day, and I thought of the bolt that long ago ran through Humberto. I thought of the countless other people in the world, from the Paleolithic Age until today, who had climbed down into caves and catacombs, tombs and tunnels, and felt the same bolt in the dark. "I had been my whole life a bell," Annie Dillard once wrote, "and never knew it until at that moment I was lifted and struck."

———

SOMETHING IN US HAS WITHERED. We in the West have
become hardened to the world, numb to certain shadowy
textures in nature, dulled to what David Abram called the
"songs, cries and gestures of the earth." Over the last so
many years, as I waded through our deepest ancestral traditions, from the Aboriginal songlines to the hidden rites of
the Magdalenians to the emergence myths of the Lakota, I
saw just how estranged we've grown from what originally
shaped us as human beings, how we've turned away from
our deepest instincts and impulses. It is in our connection
to the underground, I found, that the old ways survive. In
the subterranean dark, lost memories rumble awake. We
become raw and vulnerable, sensitive to the world's soft
enchantments, attuned to the quiet parts of our mind. We
recover our capacity to be startled and confused and left in
awe by the world. "The intake valves are open in the soul,"
as Anne Carson wrote. The underground holds the shape
of our ancestors' earliest dreams, opens us to a world that
precedes knowledge and memory—it ushers us back to the
"root of the root, the bud of the bud," in the words of E. E.
Cummings.

The underground teaches us to respect mystery. We live
in a world obsessed with illumination, where we blaze our
floodlights over every secret, strive to reveal every furrow, to
root out every last trace of darkness, as though it were a
kind of vermin. In our connection to subterranean space,
we ease our suspicion of the unknown, and recognize that
not everything should be revealed, not all the time. The un-

derground helps us accept that there will always be lacunae, always blind spots. It reminds us that we are disorderly, irrational creatures, susceptible to magical thinking and flights of dreaming and bouts of lostness, and that these are our greatest gifts. The underground reminds us of what our ancestors always knew, that there is forever power and beauty in the unspoken and unseen.

I did not come to the underground as a pilgrim. I didn't set out on any kind of mystical errand, or to retrieve sacred wisdom. But as I rummaged through the dark, I felt the world change shape around me, bending and yielding and unfolding like an enormous origami sculpture. Reality, I came to see, was more void than solid. The concrete surfaces we see and touch in everyday life are but one of many strata, all the rest of them veiled. I experienced the entire world as Steve Duncan once described New York: a giant organism, trembling and shifting, of which we can see only a sliver. Every landscape came to feel like a ghost landscape, teeming with vibrancy and potency beyond our detection. The underground helped me acknowledge the seams of ineffability in the world, teaching me to sit in peace with shadow, to embrace modes of thought that lie between the empirical and the visionary. It taught me not to shrink from the sacred, but to turn toward it, to look it full in the face. I encountered God not as a voice booming down from the clouds, but as an embrace of hiddenness, an acknowledgment of certain dark hollows whose power we will always feel, even if they may never be seen.

Today, as I move through the world, I feel the presence of the spaces beneath me, and am reminded how much of our existence remains in mystery, how much of reality continues to elude us, how much deeper our world runs beyond what we know. And from day to day, nothing leaves me so enlivened and hopeful and full of grace. The priest and ecologist Thomas Berry once wrote that as we carry through life, searching for truth and meaning in the world, "we are like a musician who faintly hears a melody deep within the mind, but not clearly enough to play it through." In the subterranean dark, I learned to listen to that faint melody—and I learned the myriad and beautiful ways in which it cannot be played.

ACKNOWLEDGMENTS

When Dante descends into the underworld, traveling down through ring after ring until he reaches the icy shores of Lake Cocytus, he is guided by the poet Virgil, without whom the journey would not be possible. In the years I spent reporting, and writing this book, I was fortunate enough to meet a whole legion of Virgils: people who guided me through strange landscapes, administered late-night pep talks, read messy drafts, and delivered bursts of inspiration large and small. Without them this book would not exist.

Thank you to the explorers, scientists, and artists who took the time to lead me underground, often into spaces that were sacred, restricted, or otherwise sensitive. Otter introduced me to the tunnel that would ultimately change the course of my life. Steve Duncan was my underground sensei in New York. He also contributed many marvelous photographs of urban subterranea, more of which can be seen at undercity.org. Russell was always up for late-night tunnel runs. Thank you to Gilles Thomas for introducing me to the Paris catacombs and, on one occasion, fighting off a subterranean smoke bomb. Luca Cuttita patiently taught me how to use a descender. The good people at SURF and Life Underground kept me safe in the depths. Sina Bear Eagle was generous with her time in the Black Hills. I am profoundly thankful to Colin Hamlett and his family, who shared their history with me and fed me kangaroo stew. Chris Nicola introduced me around and indulged my many questions about darkness. Maria Alejandra Perez helped me understand the caprices of the caver psyche. Thank you to Robert Bégouën for leading me to the bison. Many disciples of REVS left me hints and crumbs throughout the city. Thank you to Humberto, who reminisced with me. And to Holley Moyes, who helped me see the magic of the dark zone.

I am further indebted to the following people, who answered calls, fielded questions, made introductions, imparted wisdom, offered couches, or otherwise provided support: Craig Hall, Tiki Hall, Walter Tschinkel, Raina Savage, Philip Jones, Vicky Winton, Rachel Popelka-Filcoff, Ric Davies, Paul Taçon, Andreas Pastoors, Jean Clottes, Margaret Conkey, Megan Biesele, Eugenia Manzella, Moses Gates, Jazz Mandela, Liz Rush, Chris Moffett, Penelope Boston, Jan Amend, Caitlin Casar, Brittany Kruger, Duane Moser, Tom Regan, la Société française de photographie, Hatchet, Lazar, Cat, Séléna McMahan, Guillermo de Anda, Carolyn Boyd, Derek Ford, Katie Parla, "Il Papa del Sottosuolo," Henry Chalfant, Yulia Ustinova, Adriano Morabito, Emma Vaiserfirov, Moscowhite, Boris, Roman, John Longino, Berliner Unterwelten, Michel Siffre, Christian Rognant, Joshua Horowitz, Stephan Kempe, E. J. Albright, Christian Marmorstein, Jennie Schueler, Gus Jacobs, Tom and Fran Jacobs, Lina Misitzis, Natalie Reyes, Taylor Sperry, Rachel Yoder, Deke Weaver, Zhivago Duncan, Caro Clark, and Siera Dissmore.

In the writing world, I'd like to thank my agent, Stuart Krichevsky, who took me on after seeing my name mentioned parenthetically in a newspaper article and has stuck with me ever since. I appreciate his unwavering belief in this project, as well as his peerless grace and diplomacy, which smoothed out many bumps in a long road. I would also like to thank Ross Harris, Laura Usselman, and everyone who helped behind the scenes at SK Agency. I am indebted to many people at Random House, especially Julie Grau, who got this book off the ground. I appreciate, too, the work of Annie Chagnot, Mengfei Chen, and the entire production and design teams, who patiently hefted this book over the finish line. Jenny Pouech undertook the prodigious task of chasing down licenses for the book's many photographs. Samantha Weinberg, Tasha Eichenseher, and Deirdre Foley-Mendelssohn all edited portions of this book for *Intelligent Life*, *Discover*, and the *Paris Review Daily*, respectively.

This book would be a raggedy mess if not for a close circle of writers, friends, and mentors who provided heroic levels of editorial support, emotional sustenance, and cosmic benevolence from near and far. Thank you to those who taught me how to put sentences together, especially Catherine Reed, Stacie Cassarino, Chris Shaw, Suketu Mehta, Rob Boynton, Katie Roiphe, and Ted Conover. My coconspirator, Matt Wolfe, read more toxically bad drafts than anyone deserves, and always responded with patience, wisdom, and perspicacity. Rob Moor, Chris Knapp, and Elianna Kan all gave much-needed face-lifts to portions of the book, especially in the final hours. Amelia Schonbek, Nicole Pasulka, Cody Upton, Heather Rogers, and Leo Rogers edited, cheered, consoled, and rallied from within living rooms and kitchens throughout Brooklyn. Liz Flock provided whiskey-based encouragement. Ellie Ga talked me through narrative knots. Allegra Coryell propped me up, kept me sane, listened to drafts, and told me when I was being boring. Thank you.

The Institute for Public Knowledge at NYU gave me library privileges and a place to finish the book. The MacDowell Colony gave me a place to think in the woods; the New York Foundation for the Arts helped fund my research. This book would not exist in any form without the miraculous Thomas J. Watson Foundation, which sent me on my very first underground expeditions, and cracked open my relationship to the world.

Finally, a wave of gratitude and love to my family. To my sister and brother-in-law, Caroline and Tyler Ruggles, and to my nephew Henry Ruggles, who will one day read this book. To my grandmother, Carol Hunt, and to my parents, Peter and Betsy Hunt, who supported me at every turn, in every way imaginable. And thank you to Isa, who is a marvel.

IMAGE CREDITS

ABOUT THE AUTHOR

———

WILL HUNT is a nonfiction writer based in New York. He has received awards and grants from the Thomas J. Watson Foundation, the New York Foundation for the Arts, the Bread Loaf Writers' Conference, and the MacDowell Colony. He is a visiting scholar at the Institute for Public Knowledge at NYU. *Underground* is his first book.

Twitter: @willhunt__

Instagram: @willhunt__